From Creation to Chaos

From Creation to Chaos

Classic Writings in Science

Edited by
Bernard Dixon

Editorial panel

Richard Gregory, Dorothy Hodgkin,
Fred Hoyle, Peter Medawar and Jonathan Miller

Basil Blackwell

Copyright © Basil Blackwell Ltd 1989

First published 1989

Basil Blackwell Ltd
108 Cowley Road, Oxford, OX4 1JF, UK

Basil Blackwell Inc.
432 Park Avenue South, Suite 1503
New York, NY 10016, USA

British Library Cataloguing in Publication Data

From creation to chaos: classic writings in science.
1. English literature. Special subjects. Science. Anthologies
I. Dixon, Bernard, *1938–*
820.8'0356

ISBN 0–631–14976–7

Library of Congress Cataloging in Publication Data

From creation to chaos.
Bibliography: p.
Includes index.
1. Science. I. Dixon, Bernard. II. Title.
Q171.F885 1989   500   88-7971
ISBN 0–631–14976–7

Typeset in 10 on 12pt Goudy Old Style
by Columns of Reading
Printed in Great Britain by Billing & Sons Ltd, Worcester

Contents

Preface

The silliest piece of advice I ever received at school was about how to develop a more intimate acquaintance with the essence of science than was possible in the classroom or laboratory. Scrutinize intently the lives of historic greats, a kindly chemistry teacher said. Study their every move. Emulate their thinking. His suggestion sent me off in search of full-scale biographies of Newton and Einstein, Pasteur and Darwin. Abusing my library ticket, I took all four home that very afternoon and began to devour them diligently. Two weeks and many reading hours later, I was acutely depressed. The whole well-intentioned exercise had simply left me humbled at what I had learned about the cerebral exploits of my selected quartet of superhumans. How could it be otherwise for a 12-year-old trying to imagine his way into the mind of a giant – let alone four of them at once? My chemistry teacher (a man of rare ability with chalk in his hand) really should have known that such an experience was as likely to engender gloomy pessimism as inspirational uplift.

Much the same applies to the art of writing. How on each can a tongue-tied adolescent be tutored in the syntax, words and cadences of William Blake, Dylan Thomas, Ernest Hemingway or Robert Burns without concluding that theirs is a craft which it would be prudent to avoid? This is the very process which convinces many otherwise intelligent and sensitive individuals that they 'cannot write'. Indeed, lots of well-meaning teachers of Eng. Lit. and Eng. Lang. could well be contributing to the illiteracy they deplore, by brandishing before their pupils models of unattainable excellence. They have the best of motives, without doubt. And I certainly do not argue against youngsters reading Blake, Thomas, Hemingway and Burns for simple pleasure. Quite the contrary. What strikes me as odd, however, is the notion of any of us trying to match, word for word, some of the finest artistic outpourings in the history of the world.

I have two counter-suggestions for all youngsters and oldsters stuck with an impression of writing as an activity, like surf-boarding, at which they will never acquire even modest competence. The first piece of advice is to learn from *bad* writing. Whenever we find ourselves tossing aside a newspaper story or magazine article, intending to finish it later, it's worth considering why we are doing so. The reason may be tea-time, bed-time or a call of nature. But as often as not the explanation can be found in the text itself, which is simply not enticing enough to sustain our interest. The prose may be dourly passive. It may meander without direction or purpose. We may have run into gratuitous jargon, or ground to a halt amidst thickets of impenetrable syntax. Whatever the explanation, there are likely to be lessons which we can turn to advantage in our own writing.

The tactic is dreadfully well illustrated by a booklet which arrived on my desk a few years ago. Although its purpose was to help people write technical reports, the document itself contained some of the most deathly prose I had encountered in a very long while. Try these two sentences:

> Information workers (librarians, information scientists, archivists, indexers, bibliographers, database managers, information officers – and all the other titles by which such workers may be known), as well as those training to be information workers, are required to produce reports for many different purposes and within a variety of subject environments. This publication sets out only to suggest guidelines for the production of reports in those information-oriented environments; and in most cases the principles are the same – no matter what the context – but a particular situation may demand a certain emphasis, selected set of actions or specific content features.

Those sentences are tedious, over-long, verbose, clotted, boring and unclear. Because they are so grotesque, however, the stylistic faults come across with considerable power. We can learn important lessons by inspecting such specimens, just as pathologists learn from even the most unattractive objects and tissues that arrive in their laboratories.

My second counter-suggestion is indeed to read great prose – but not in any attempt at conscious imitation. A far better and more practical motive is sheer, self-indulgent pleasure. Adopt that perspective and the lessons of style, clarity, rhythm, balance and syntax will sink in by subconscious osmosis. Which brings me to science and to this anthology. Over the past five years, with the help of friends, colleagues and correspondents, I have been collecting samples from the literature of science. The material comes from the pens and word processors of both scientists and science writers. It ranges from ancient to modern, and from textbooks to popular magazines in original sources, and has been selected solely on the basis of literary quality. I hope that the resulting collection will help to demolish the well-worn cliché that scientists, by their nature and training, cannot write. Its sole *raison d'être* is to provide pleasure for the reader. I cannot, of course, predict whether the book will also be brandished as a didactic tool by information workers (and all the other titles by which such workers may be known).

For their suggestions and support in this project, my thanks go to Sir Walter Bodmer, Ben Bova, Dr Jon Darius, Dr Richard Dawkins, Dr John Durant, Sir Alan Cottrell, Margaret Galbreath, Dr Geoff Holister, the late

Professor John Humphrey, Professor R. V. Jones, Kath Adams, Dr Bryan Large, Dr John Laurent, Dr Stephen Lock, Dr D. J. Mabberley, Professor Jack Meadows, Patrick Moore, Kim Pickin, Sir George Porter, Dr Michael Rodgers and Dr Tony Smith. I am particularly grateful to my fellow editors Professor Richard Gregory, Professor Dorothy Hodgkin, Sir Fred Hoyle and Dr Jonathan Miller. Above all, I feel privileged to have received the active support of the late Sir Peter Medawar, whose distinctive elegance in science, literature and philosophy may never be equalled.

P. W. Atkins

An extract from 'Why things change', a chapter of *The Creation*, 1981

At its most rudimentary, a chemical reaction is a rearrangement of atoms. Atoms in one arrangement constitute one species of molecule, and atoms in another, perhaps with additions or deletions, constitute another. In some reactions a molecule merely changes its shape; in some, a molecule adopts the atoms provided by another, incorporates them, and attains a more complex structure. In others, a complex molecule is itself eaten, either wholly or in part, and becomes the source of atoms for another.

Molecules have no inclination to react, and none to remain unreacted. There is, of course, no such thing as motive and purpose at this level of behaviour. Why, then, do reactions occur? At this level too, therefore, there can be no motive or purpose in love or war. Why then do they occur?

A reaction tends to occur if in the process energy is degraded into a more dispersed, more chaotic form. Every arrangement of atoms, every molecule, is constantly subject to the tendency to lose energy as jostling carries it away to the surroundings. If a cluster of atoms happens by chance to wander into an arrangement that corresponds to a new molecule, that transient arrangement may suddenly be frozen into permanence as the energy released leaps away. Chemical reactions are transformations by misadventure.

Atoms are only loosely structured into molecules, and explorations of rearrangements resulting in reactions are commonplace. That is one reason why consciousness had already emerged from the inanimate matter of the original creation. If atoms had been as strongly bound as nuclei, the initial primitive form of matter would have been locked into permanence, and the universe would have died before it awoke.

The frailty of molecules, though, raises questions. Why has the universe not already collapsed into unreactive slime? If molecules were free to react each time they touched a neighbour, the potential of the world for change would have been realized long ago. Events would have taken place so haphazardly and rapidly that the rich attributes of the world, like life and its own self-awareness, would not have had time to grow.

The emergence of consciousness, like the unfolding of a leaf, relies upon restraint. Richness, the richness of the perceived world and the richness of the imagined worlds of literature and art – the human spirit – is the consequence of controlled, not precipitate, collapse.

Energy itself holds the key to its own degradation. Molecules have the opportunity to react when they meet, but they actually do so only if their atoms are loose enough to wander into new arrangements and to expose themselves to opportunities for misadventure. While frail, molecules are not floppy.

In order to explore, the atoms of molecules must be marginally loosened. They are loosened if energy enters the molecule and stimulates vibration, because a vigorously vibrating molecule is a loose cluster of atoms. And how does the energy enter the molecule? It enters by chance. By chance energy may jostle its way in and be there at the moment the molecules happen to meet. By chance a pair of molecules may meet while they happen to be favoured with more than the average share of energy. Then their atoms may wander, and by wandering, react.

I should like to pause at this point to summarize the argument so far. We have seen, at a molecular level at least, how chaos both drives and restrains the world. Collapse into chaos motivates change, for all natural events are outcomes of the tendency to dispersal. Chaos also stabilizes form, because the chance is only small that molecules are favoured with enough energy for them to explore possible alternative arrangements. We are both led on and held back by chaos: chaos is both the carrot and the cart.

If everything, both structure and change, is the outcome of chance orchestrations of chaos, there must be chains linking the superficial to the deep. I should like to try to indicate their nature, if not their detailed form.

Evolution is reaction by seduction. Complex molecules can acquire even greater complexity in stages instead of attempting a single grand passion. One molecule may be able to discard a few atoms to a congenial partner, pick up a few others elsewhere, and in due course chance upon a destination. Only a little reorganization has to take place at each step, and so only a little loosening is required for each one. Since small chance influxes and abundances of energy are more likely to occur than big ones, the overall process may occur much more quickly than if enough energy had to arrive for there to be reaction in a single stride. That is reaction by multiple misadventure, reaction down the slippery slope. Whether or not the reaction can proceed then becomes mainly a matter of logistics, or the supply of little molecules at the appropriate time in the meal.

The whole course of evolution can be regarded as a geared and cooperative dissipation of energy. Every stage of evolution, including the steps that gave rise to complex molecules out of simpler ones, to people out of slime, and the processes involved when species are confronted with competition, proceeds by dissipation.

Molecules did not aim at reproduction: they stumbled upon it. Accretion of complexity reached the point where one molecule was so structured that the sequence of reactions it could undergo, under the casual pressure of dispersal, led by chance to the formation of a replica. That molecule naturally had the same reproductive ability, and even though the first might have been eliminated by the evaporation of a pond, the daughter could continue the line. At every stage of replication there was opportunity for modification because slightly different smaller molecules were in the vicinity and could be incorporated. Many of these daughters may have been unviable, or less successful at replication than their ancestors and sisters; but some were more successful, and flourished into elephants.

Perceptions of the external world developed in subtlety with the evolution of the complexity of the body. Those perceptions, like decisions to embark upon activities and our reflections on our own activities and those of others, are all manifestations of reactions. We interact with the external world as the breeze of events shifts specially responsive groups of atoms fashioned into eyes and ears.

Since reactions are aspects of chaos, perceptions, decisions, and reflections are also ultimately driven by an underlying tendency to chaos. The apparent complexity of consciousness is the outcome of the complexity of the interdependence of the reactions geared to this decay, and there is no need to regard it as an intrinsic complexity embellished by a soul. Behaviour is the complex organization of simple processes, and the complex structure of the brain is the complex gearing that marshals simplicity into apparent complexity. The structure ensures that simple chemical processes within the cells of the brain are coordinated into a whole which is both sufficiently complicated to be rich in properties and sufficiently unpredictable to encompass imagination and invention.

Take perceiving. Its essence is the acquisition of information about events external to the brain's bearer, and events not wholly external, as in pain. Bodies have antennae – nerve endings – that respond to their environment and capture information. These sensors, bunched into things like eyes, trigger signals to the brain. In vision, for instance, a molecule in an eye is struck by light, uncoils, and no longer fits its original slot. The light brings energy which loosens the atoms. The atoms ramble, and in the course of rambling their energy jostles away. The molecule remains frozen in its new and now incompatible shape. The ejection of the molecule from its slot lets another molecule change its shape, which triggers another reaction. That reaction triggers a pulse of current along the nerve to the brain itself. The nerve ramifies, the pulse is spread to a multitude of cells within the brain, and in each one its arrival results in a chemical modification. The cells' constitutions determine how they respond to future pulses, and whether they send new pulses down some channels or down others. And in due course, but perhaps not for a decade, the perception of an event influences a deed.

Every process in this chain is propagated from stage to stage through the motiveless agency of chaotic dispersal. The light loosens the molecule, which then uncoils by misadventure. The molecule is ejected because it has freedom to roam, and energy to lose. A reaction takes place when the ejection of the molecule allows the one remaining to explore new arrangements. The electric pulse is squeezed along the nerve by a sequence of reactions, each one triggered by its neighbour, and each trigger permitting molecules to wander into new arrangements. The chemical reactions in the ramification of nerve cells in the brain are similarly triggered, so is the deployment of the current as it rings for years through the brain. All the processes in the sequence, up to, including, and extending beyond the subsequent deed, are driven forward by the chaos they unleash. That subsequently we laugh or cry, or in due course love, argue, or despair, is determined by the long and complex history of events, all driven by dispersal.

I find it perplexing that to some, even now, it appears that the richness of the brain's properties, properties like perceiving, remembering, acting, deciding, and inventing, cannot have emerged by itself, or that such richness cannot be the outward display of inner motivelessness. It is so important to see through the illusion of complexity into the simplicity underneath. Of course, we might not be able to trace the simple steps that constitute a perception or an opinion or precede or bring about an action; but underneath there is no doubt that they are there. Yet I would not wish this view to be taken as an elimination of the wonder of life: it should, though, redirect the wonder. What wonder there is, should, in my view, not be at the benevolence and subtlety of external intervention, for that leads to the unnecessary intrusion of a spirit and the invention of a soul. It should instead be wonder at the realization that underlying simplicity can have such glorious manifestations when elaborately coordinated, and that such coordination can grow through the selection of evolution. The only immortal soul man has is the lasting impression he makes on other men's minds.

We do not see one thing and die. The body must be rewound to respond again, and nerves have to be prepared to propagate again. Every step in perception and action is reaction, and every reaction can be undone. A suitable reagent has to be supplied which has the energy to permit yet another haphazard exploration of arrangements, and to induce the molecules to re-form their earlier structures. In other words, we have to eat.

We have seen that perception and action are both powered by the tendency of energy towards chaos. Both degrade the quality of energy in the universe, and both ultimately involve energy's corruption. In eating we import more high-quality energy from our surroundings, and recharge our bodies by letting it disperse into our cells, where it is ready for the next step of degradation, such as an act of perception, of action, or of invention. Every action is corruption; and every restoration contributes to degradation.

At the deepest level, decisions are adjustments of the dispositions of atoms in the molecules inside large numbers of cells in the brain. The underlying reasons for those changes are the same as for any process. The atoms have no will to move, but given an opportunity they explore, with the risk of being trapped when the energy jostles into the world and dissipates. Every change in the complexion of the cells and their interconnections is at heart brought about by a natural disposition to chaos. That this motiveless, purposeless, mindless, activity emerges into the world as motive and purpose, and constitutes a mind, is wholly due to the complexity of its organization. As symphonies are ultimately coordinated motions of atoms, so consciousness emerges from chaos.

Decisions are taken on the basis of the predisposition of the brain. The manner in which chaos emerges into the world to take the name of action depends upon the state of preparation of its cells. The consequences of changing the state of a single cell depend upon the state already existing in the cells with which it is in contact. So the whole of our personal history, so long as our cells survive, channels the ramifications of chaos. That cells switch the activity of the brain to some cells and not to others, and in the process become modified so that subsequent pulses, perhaps returning from those recently stimulated, are channelled elsewhere, is the complexity of organization that feeding on chaos marshals it into coherence.

Inheritance, genetic information transmitted by reactions, lays down the structure of the brain and imposes a pattern of switches. Experience, the lifelong sequence of reactions responding to influence, then ceaselessly modifies and develops the network. Age is the death of cells and consequently the loss of subtlety. Senility is the decay of the sophistication of the organization of the circuitry, and the consequent failure of the brain to coordinate underlying chaos into brilliance.

So long as we can restore our cells by hunting for high quality, undispersed energy in the outside world, transferring some of it to our cells, then so long can our complexity be sustained. A bleak yet honest view is that living is therefore a struggle (a struggle ultimately driven not by purpose but by dispersal) to discard low-quality energy into the surroundings and to absorb high-quality energy from them. In a sense, we corrupt the outside world in order to have an inner life. The chain of consumption, men eating cows, cows eating grass, grass eating mountains and living off the sun, is what has grown up through evolution as an interlocked mechanism of dispersal. There is no need to look for a purpose behind it all: energy has just gone on spreading, and the spreading has happened to generate elephants and enthralling opinions.

I should like to add a postscript. The singular property of the brain is that its response to circumstances is to a degree under its own control. It can take advantage of opportunities to select paths towards its own annihilation, as in despair or an inclination to martyrdom. Or it can take advantage of opportunities to select paths towards the fulfilment of its potential, as in acts of comprehension and creation. These inclinations are consequences of the pre-existing state of the brain, its chemical composition when the thought or inclination emerges and is rendered into action. Free will is merely the ability to decide, and the ability to decide is nothing other than the organized interplay of shifts of atoms responding to freedom as chance first endows them with energy to explore, and then traps them in new arrangements as their energy leaps naturally and randomly away. Even free will is ultimately corruption.

Francis Bacon

Two extracts from Novum Organum, 1620

Lastly, let none be alarmed at the objection of the Arts and Sciences becoming depraved to malevolent or luxurious purposes and the like, for the same can be said of every worldly Good; Talent, Courage, Strength, Beauty, Riches, Light itself, and the rest. Only let mankind regain their rights over Nature, assigned to them by the gift of God, and obtain that power, whose exercise will be governed by right Reason and true Religion. . . .

Thus (as we have before observed), had any one meditated on balistic Machines, and Battering Rams, as they were used by the Ancients, whatever application he might have exerted, and though he might have consumed a whole life in the pursuit, yet would he never have hit upon the Invention of Flaming Engines, acting by means of Gunpowder: nor would any person, who had made woollen Manufactories and Cotton the subject of his observation and reflection, have ever discovered thereby the nature of the Silk-worm or of Silk.

Hence all the most Noble Discoveries have (if you observe) come to light, not by any gradual improvement and Extension of the arts, but merely by Chance; whilst nothing imitates or anticipates Chance (which is wont to act at intervals of ages) but the Invention of Forms.

There is no necessity for adducing any particular Examples of these Instances, since they are abundant. The plan to be pursued is this; all the Mechanical, and even the Liberal Arts (as far as they are Practical), should be visited and thoroughly examined, and thence there should be formed a Compilation or Particular History, of the Great Masterpieces, or most finished Works in each, as well as of the Mode of carrying them into effect. . . .

Margaret A. Boden

'In search of unicorns', an article from *The Sciences*, 1983

Only Princess Diana's wedding dress was awaited more impatiently, greeted more enthusiastically, and copied more slavishly than are new ideas in psychology. Because psychology lacks a generally accepted theoretical uniform that fits all figures and pleases all tastes, it is especially prone to changing fashions. Psychologists do not even agree about what basic items their science's wardrobe should contain – what the right questions are to ask. While some psychologists may be confident that they are posing the central questions, others will surely disagree. So from the unadorned statement that someone is a psychologist, one can infer very little about that person's professional beliefs, or even interests.

Given this disagreement over what style of theorizing best suits the mind, any new approach is likely to be hailed as the missing paradigm, the link transporting psychology from myth to science. The computational approach – in which minds are compared to computer programs – is the most recent psychological fashion. But it is not the first intellectual style to be welcomed as psychology's savior, nor is it the first to be mocked by those preferring different modes.

Distinct psychological fashions have been designed by such theorists as Sigmund Freud, Ivan Pavlov, Jean Piaget, B. F. Skinner, and R. D. Laing, but none of these has achieved the enduring status of a Chanel suit. Their popularity has waxed and waned over the years and varies among different groups. Workaday styles in psychology – such as intelligence tests and personality profiles – have been widely adopted for practical purposes, but many see them as disguising the true nature of what they are intended to display. And if we look to psychological accessories, whose designers are such fringe figures as Wilhelm Reich, of the Orgone Box, Werner Erhardt, of Erhardt Seminar Training, and Carl Janov, of the Primal Scream, it seems that no psychologist can be so maverick as to lack a body of faithful followers, while none is so authoritative as to persuade all comers of his or her theoretical infallibility. Psychology is not a unified church.

But, unified or not, it is a church. The different styles of psychology resemble religious sects, arousing emotional commitment and antagonism to a degree rarely seen in other branches of scientific inquiry. This is not surprising, for any denomination in the field – whether venerated as the science of mind, brain, or behavior – has implications that bear on deep issues of self and society. So psychological theories typically arouse not only intellectual disagreement and rejection but also passionate denouncement and scathing ridicule.

As in more theological forms of sectarianism, psychologists will go to great lengths, or sink to surprising depths, in opposing those theoretical fashions they find unattractive. Even in the gentlemanly nineteenth century, the psychologist and philosopher William James remarked of the new, experimental statistical psychology that it 'could hardly have arisen in a country whose natives could be *bored*. Such Germans as Weber, Fechner, and Wundt obviously cannot.' By the 1920s the invective had intensified. J. B. Watson's brainchild, behaviorism, had conquered the American academies within a few years of its birth, in 1913. E. C. Tolman sneeringly described it as 'mere Muscle Twitchism', and William McDougall as 'a most misshapen and beggarly dwarf' – a description that, in the world of *haute couture*, would be damaging indeed. More recently, Noam Chomsky has ridiculed the preeminent behaviorist Skinner by saying that, according to Skinner's views on reinforcement, the best way to encourage an artist would be to stand in front of his paintings yelling 'Beautiful!' at the top of one's voice.

But behaviorists are not the only theorists to be attacked. Freud himself has been accused of systematic intellectual dishonesty by the English philosopher Frank Cioffi, and the German psychologist Hans Eysenck has viewed Freudian theory as a classic case of the Emperor's new clothes, mocking it with a suitably italicized account of a railway journey: 'the train *enters a dark tunnel*', the signal arms '*rise* as we approach *and fall* again as we pass', and we '*sharpen a pencil*' to write a postcard, but – horror of horrors – 'the *point drops off*.'

Borrowing the words of the seventeenth-century philosopher Thomas Hobbes, Guy Robinson has applied perhaps the most scathing dismissal – 'When men write whole volumes of such stuff, are they not mad, or seek to make others so?' – to the most recent psychological fashion. This is the computational approach, whose disciples describe the mind and thinking with concepts drawn from artificial intelligence. This involves the use of computer programming to study the structure and function of knowledge. But unlike programs for calculating tax rebates or matching computer dates, which are rigid and inflexible, work in artificial intelligence focuses on *intelligent* information-processing abilities, which enable a system to cope flexibly with changing and largely unknown situations. The programs specify computations that enable computers to converse (by teletype) in natural language; understand spoken speech; recognize objects seen in widely varying positions or lighting conditions; plan complex tasks involving unpredictable conditions; make sensible guesses when specific knowledge is not available, and the like.

Computation in this sense does not mean mere counting, but *any* symbolic process of inference, comparison, or association. The symbolism may be numerical (for counting is one example of computation), or of some other form, such as verbal, visual, or logical. Seen from this viewpoint, the mind is a system for manipulating symbols. It contains many representations of aspects of the world (and other possible worlds), and a variety of rules for building, changing, comparing, and drawing inferences. Psychological questions, accordingly, concern the structure and content of mental representations and the ways in which they can be generated, augmented, and transformed. Thought, experience, and motivation – and the myriad differences among individual people that lead to the fascinating human pastime of gossip – are grounded in computational processes.

In the Middle Ages, for instance, the thoughts and actions of people who set off in search of unicorns, expecting to find them in the forest with their heads resting in the laps of virgins, were guided by a specific mental representation – the goal of finding a unicorn in those circumstances. We in the twentieth century can form similar representations; that is, we can think about medieval beliefs and ideas about unicorns. But we do not, as a consequence, guide our footsteps into the forest, because our minds represent the notion that unicorns do not exist. An essential precondition for purposeful, voluntary action – that the goal be believed, rightly or wrongly, to be at least *possibly* attainable – is thus not satisfied.

Suppose we were to suspend our disbelief in unicorns, or to discount it, and then to venture into the woods for a fanciful picnic, dressed appropriately in tunic and hose and carrying a silken halter. This would require a temporary transformation of our representation of unicorns so that their nonexistence was either not recognized, or else not allowed to veto the afternoon's plans. That is, the check on whether the plans were realistic would not be carried out, so that from the judgment 'there are no unicorns' we would not draw the inference that 'there is no point in forming the goal of finding a unicorn.' Searching for unicorns, then – and also refusing or pretending to do so – are human activities that depend on the functioning of specific rules and representations in the mind. If these are transformed, by learning, reasoning, or fancy, then the person's thoughts and actions relating to unicorns will be different.

These differences in behavior and experience may be subtle or coarse-grained, for the representations concerned are varied and complex. Planning a unicorn-hunt, whether for fun or for real, requires that our minds contain more than the concept of unicorn. We must also understand the concepts of virgin and of forest, and be able to represent their probable locations and recognize them when we get there. We must be able to plan how to reach the forest, and how to creep up on a virgin and a unicorn without frightening either. And if we hope to catch the unicorn, we must not forget the halter. If we cannot get one of silk, would a hemp one do instead? According to our current sensibilities, it probably would. But according to an older, magical, viewpoint, it might not.

The psychological interest here is not in which conception of unicorn-hunting is true and which is false, but in how people can make such mental representations and be guided by them, regardless of whether they are realistic. Understanding how something is possible is more important, in this sense, than predicting what will actually happen. Of course, if someone believes a unicorn to be a sea-creature, half fish and half woman, we can predict that he will not search for unicorns in the forest. But the theoretical interest is in how the familiar concept of unicorn can be integrated with a person's powers of perception, planning, and persuasion, so as to generate a unicorn-hunt. This integration may be extremely complex, involving comparisons of priorities (What else might one do this afternoon?) and individual life-styles (Which of one's friends would appreciate the enterprise?).

Even in the Middle Ages, though, life was not focused solely on unicorns. And today people solve problems of varying sorts – from cooking a meal to designing motor-bikes to writing sonnets. Similarly, people hold beliefs of different sorts about different kinds of things, beliefs that are largely idiosyncratic and not always consistent with each other. As Walt Whitman said, 'Do I contradict myself? Very well then I contradict myself (I am large, I contain multitudes).' An adequate theoretical psychology should help explain how all these problems and beliefs can coexist in an individual mind. It should address such fundamental questions as: How do people recognize different types of problems and classify some as tractable and others as hopelessly beyond grasp? What mental processes enable us to build or acquire our various beliefs? How do we interrelate them, by inferring one belief from another or by recognizing inconsistencies? If we decide not to tolerate an inconsistency, how do we transform our minds' content or organization, or both, accordingly? For example, how do we relate evolutionary biology and Christianity? Can they be mental bedfellows, and if so what sort of conceptual bolster might be needed to be put down the middle of the bed?

No one, at present, can answer all these questions. But they are the sorts of problems to which computational psychology is especially well suited, for they concern the ways in which we store, retrieve, compare, and transform various sorts of symbolically represented information. Indeed, psychological questions in general – whether they concern belief, problem-solving, purpose, choice, language, perception, memory, or even emotion – can be understood as computational questions about mental rules and representations.

A connoisseur of the history of psychological fashions might observe that the newly arrived computationalists are not the first to style the mind as a domain of symbolic representation and transformation. Freud, for example, thought of defense mechanisms as involving different sorts of psychological transformation. Introjection and displacement, for instance, transform the object of neurotic hatred in distinct ways: introjection shifts it from another person to oneself; displacement to some third person, conceived of as somehow analogous to – symbolic of – the original. Similarly, dreams and slips of the tongue involve strings of symbolic transformations, which depend on idiosyncratic associations and comparisons as well as on generally interpretable symbols (such as tunnels). But Freud's ideas in this regard, like those of other noncomputational psychologists, are suggestive rather than specific, vaguely expressed rather than rigorously defined.

The computational approach, by contrast, offers precisely definable concepts, because a program has to be expressed clearly, as a set of instructions for specific symbol manipulations, if a computer is to accept it. If a program is written in some high-level programming language (one that more closely resembles human language than the series of binary digits that the computer will ultimately act on), the programmer can ignore the more basic operations involved – much as we may think of a task in terms of some high-level goal, being unable to specify the details of how we tackle it. But, since clarity is essential, work in artificial intelligence provides a rich source of clear distinctions between many types of symbolic representation and interpretative process.

Moreover, this style of theorizing highlights process as well as structure, since a program has to tell the computer not only what to produce but also how to produce it. Non-computational psychologists often take psychological change for granted, assuming that it can be specified sufficiently by stating the initial and final mental states involved. However, the process of mental transformation is itself problematic. In programming, a failure to suggest a way in which a change might be effected will show up as a glaring gap over which the uninstructed computer cannot leap. Some computational account of how to make the leap must be supplied if the program is to function. In short, the pictures of the mind that are drawn in the computational style are more like movies than pinups.

Think, for example, what happens in our minds when we discern what someone means by the word *it* on any particular occasion. Consider this snatch of conversation from *Alice's Adventures in Wonderland*:

> 'Even Stigand, the patriotic Archbishop of Canterbury, found it advisable –'
> 'Found what?' asked the duck.
> 'Found *it*,' replied the mouse, rather crossly. 'Of course you know what "it" means?'
> 'I know what "it" means when *I* find a thing,' said the duck. 'It's generally a frog, or a worm. The question is, what did the Archbishop find?'

Obviously, in the duck's last remark, 'it' refers to the object found by the duck, whereas in the first sentence of the exchange 'it' does not refer to an object at all (which is why the question here is *not* 'what did the Archbishop find?'). Stating the grammatical principles involved here (so as to say precisely what was the duck's mistake in asking *what* the Archbishop had found) is very difficult. And it is well-nigh impossible to suggest a series of psychological processes for interpreting the word *it* in its various uses (there have been three in this sentence so far!), processes that might explain what goes on in our minds when we understand everyday language. Or, rather, it is well-nigh impossible without the discipline of programming.

A common criticism of the computational approach to psychology is that while a computer program can achieve a certain result, such as recognizing a unicorn, playing chess, or interpreting the word *it*, people do not necessarily reach that result in the same way. While there are many different levels at which one might specify the way in which a program does something – it might do it in the same way as people under one description, but in a different way under another – one cannot pass directly from computation in a program to thinking in a person. Still, even psychologists who see the computational approach as radically misconceived often admit that it may be scientifically useful for generating hypotheses. Thus, many who doubt that computational methodology will answer all their questions are prepared to use it until its limits can be established.

The newly fashionable talk of computers and programs, however, is not acceptable in all salons, for many people see it as the 'punk-gear' of psychology, as an aggressive rejection of traditional styles. From this viewpoint, the computational approach appears to offer a chilling picture of humanity that is not only false but also dangerously dehumanizing. It has been criticized as an obscene and deeply humiliating view, one that will deaden our personal responses and our valuation of purpose, desire, and emotional life. It is bad enough, such critics complain, to say (with Freud) that we are driven by irrational drives and uncontrollable anxieties, or (with Skinner) that, like rats or pigeons, we are slaves responsive only to environmental conditioning. But to put people on a par with computers is even worse than bringing us down to the level of unreasoning beasts. Little wonder, then, that such humanists accuse proponents of the computational approach of being 'mad', or, worse, of seeking to make others so.

Like beauty, however, madness may be in the eye of the beholder. Fears of the computational approach are mistaken. They rest on the failure to realize that describing something (whether person or computer) as a symbol-manipulating system is conceptually quite distinct from describing the physical hardware that embodies the computational powers concerned. The first description requires computational ideas, whereas the latter employs the terms of physics, chemistry, and physiology. Computational psychology does not support the mechanization of the world picture that has been brought about by the natural sciences, and by such 'scientific' styles of psychology as behaviorism. Rather, it emphasizes the richness and subtlety of our mental powers, a richness that has often been intuitively glimpsed, at least by poets and novelists, but never theoretically recognized by psychologists. It admits the influence on our lives of shared cultural beliefs, of individual ideas, interests, purposes, and choice, and of self-reference and self-knowledge. And it provides rigorous hypotheses about the mental processes that underlie such influences and make them possible.

But, the power of fashion in psychology being so great, perhaps the computational style is a mere passing fancy? Is it a trendy fad born of a technological society, doomed to obsolescence because of its tin-can irrelevance to human realities? Or is it a classic, enduring contribution, the seed of the long-awaited general paradigm of psychology? Its being currently fashionable need not debar it from that role, for although fashion is largely ephemeral, some modes last. While hats and halter-necks may be in or out, and colors change from season to season, shoes endure, despite changes in their details. How could they not, being so useful to soft-footed walking creatures?

The computational style in psychology will likewise survive, for it is so well suited to the representational anatomy of our minds. It offers a lasting insight into important mental features that other psychological approaches have recognized less clearly, or wholly ignored. It illuminates not only our cognitive or intellectual powers, but also our capacities for purposeful action and moral choice. It will change, to be sure, and some of its changes will doubtless be as shocking as next month's cover of *Vogue*. Many of the currently favored types of computation will be superseded by others; even today, different types are preferred by different theorists. But the computational approach will endure, for it has provided a standard of rigor and clarity that must make us permanently dissatisfied with less.

William Boyd

An extract from 'The viral pneumonias', from *Pathology for the Physician*, 1958

At the height of the pandemic there was a more diffuse and general involvement of the lung, due either to the heightened virulence of the primary invader or to the action of secondary invaders. The lungs were voluminous and covered by a fibrinous exudate. They had a firm or rubbery feel, and areas of consolidation could be felt in every lobe. Excised portions when placed in water floated a short distance below the surface, but did not sink to the bottom as in lobar pneumonia. Not only were they red and congested, but numerous minute hemorrhages were scattered over the surface. The cut surface was a vivid red, with here and there splashes of a darker color. The lung was wet and waterlogged, and the hemorrhagic edema was at once evident, for blood-stained fluid poured from the surface and collected in the basin in which the organ was lying. The bronchioles were no longer filled with frankly purulent material which could readily be squeezed out, but with a more watery secretion, indicating an intensely virulent action on the part of the irritant. The consolidated areas were often large and confluent, lobular pneumonia was common, and even a lobar consolidation might be simulated. The bronchial glands were enlarged, congested and edematous. The pleural surface was covered by a fibrinous exudate and the pleural cavity usually contained a small quantity of thin watery fluid, clear, slightly turbid, or it might be blood-stained. Empyema was quite uncommon in pure infections with Pfeiffer's bacillus. Its presence practically always indicated a secondary invasion by pneumococci, streptococci or staphylococci. . . .

Vannevar Bush

Extracts from *Science Is Not Enough*, 1967

In these circumstances it is not at all strange that the workers sometimes proceed in erratic ways. There are those who are quite content, given a few tools, to dig away, unearthing odd blocks, piling them up in the view of fellow workers, and apparently not caring whether they fit anywhere or not. Unfortunately there are also those who watch carefully until some industrious group digs out a particularly ornamental block, whereupon they fit it in place with much gusto and bow to the crowd. Some groups do not dig at all, but spend all their time arguing as to the exact arrangement of a cornice or an abutment. Some spend all their days trying to pull down a block or two that a rival has put in place. Some, indeed, neither dig nor argue, but go along with the crowd, scratch here and there, and enjoy the scenery. Some sit by and give advice, and some just sit. . . .

Science has a simple faith, which transcends utility. Nearly all men of science, all men of learning for that matter, and men of simple ways too, have it in some form and in some degree. It is the faith that it is the privilege of man to learn to understand, and that this is his mission.

If we abandon that mission under stress we shall abandon it forever, for stress will not cease. Knowledge for the sake of understanding, not merely to prevail, that is the essence of our being. None can define its limits, or set its ultimate boundaries. . . .

Nigel Calder

An extract from *Timescale: An Atlas of the Fourth Dimension*, 1983

In caricatures of Mother Earth, where land masses ran riot, the first lords were colored slime, then upstart worms, then sprawling mammal-like reptiles, all for far longer intervals than humans have existed. In Darwin's paleontological estimation, ours is not so much the era of the risen ape as the Age of Barnacles. Holy mountains turn out to be wreckage of continental traffic accidents, while Chicago and Leningrad sit in the chairs of glaciers gone for lunch. All in all, the refurbished creation myth owes more to Groucho than to Karl Marx. It is a tale of hungry molecules making dinosaurs and remodeling them as ducks; also of cowboys who put to sea, quelled the world with a magnetic needle, and then wagered their genes against a mushroom cloud that knowledge was a Good Thing.

Rachel Carson

Part of 'A fable for tomorrow', the first chapter of *Silent Spring*, 1962

There was once a town in the heart of America where all life seemed to live in harmony with its surroundings. The town lay in the midst of a checkerboard of prosperous farms, with fields of grain and hillsides of orchards where, in spring, white clouds of bloom drifted above the green fields. In autumn, oak and maple and birch set up a blaze of colour that flamed and flickered across a backdrop of pines. Then foxes barked in the hills and deer silently crossed the fields, half hidden in the mists of the autumn mornings.

Along the roads, laurel, viburnum and alder, great ferns and wildflowers delighted the traveller's eye through much of the year. Even in winter the roadsides were places of beauty, where countless birds came to feed on the berries and on the seed heads of the dried weeds rising above the snow. The countryside was, in fact, famous for the abundance and variety of its bird life, and when the flood of migrants was pouring through in spring and autumn people travelled from great distances to observe them. Others came to fish the streams, which flowed clear and cold out of the hills and contained shady pools where trout lay. So it had been from the days many years ago when the first settlers raised their houses, sank their wells, and built their barns.

Then a strange blight crept over the area and everything began to change. Some evil spell had settled on the community: mysterious maladies swept the flocks of chickens; the cattle and sheep sickened and died. Everywhere was a shadow of death. The farmers spoke of much illness among their families. In the town the doctors had become more and more puzzled by new kinds of sickness appearing among their patients. There had been several sudden and unexplained deaths, not only among adults but even among children, who would be stricken suddenly while at play and die within a few hours.

There was a strange stillness. The birds, for example – where had they gone? Many people spoke of them, puzzled and disturbed. The feeding stations in the backyards were deserted. The few birds seen anywhere were moribund; they trembled violently and could not fly. It was a spring without voices. On the mornings that had once throbbed with the dawn chorus of robins, catbirds, doves, jays, wrens, and scores of other bird voices there was now no sound; only silence lay over the fields and woods and marsh.

On the farms the hens brooded, but no chicks hatched. The farmers complained that they were unable to raise any pigs – the litters were small and the young survived only a few days. The apple trees were coming into bloom but no bees droned among the blossoms, so there was no pollination and there would be no fruit.

The roadsides, once so attractive, were now lined with browned and withered vegetation as though swept by fire. These, too, were silent, deserted by all living things. Even the streams were now lifeless. Anglers no longer visited them, for all the fish had died.

In the gutters under the eaves and between the shingles of the roofs, a white granular powder still showed a few patches; some weeks before it had fallen like snow upon the roofs and the lawns, the fields and streams.

No witchcraft, no enemy action had silenced the rebirth of new life in this stricken world. The people had done it themselves.

This town does not actually exist, but it might easily have a thousand counterparts in America or elsewhere in the world. I know of no community that has experienced all the misfortunes I describe. Yet every one of these disasters has actually happened somewhere, and many real communities have already suffered a substantial number of them. A grim spectre has crept upon us almost unnoticed, and this imagined tragedy may easily become a stark reality we all shall know.

What has already silenced the voices of spring in countless towns in America? [*Silent Spring*] is an attempt to explain.

Lord Cherwell

Extracts from 'The importance of technology and fundamental research to industry', from *Chemistry and Industry*, 1954

I think the main reason why the British chemical industry has been able to develop so rapidly and to overcome the handicaps to which I shall revert presently, is the recognition by the great chemical firms in this country that fundamental research is essential to progress. More perhaps than in any other industry have you shown by work in your own laboratories and by the help you have given to people working in university laboratories that you realise the vital importance of pure fundamental investigation which at first sight may seem to have no relation whatever with practical applications. I do not think there is any other industry in this country that has shown so much foresight and understanding.

But the British chemical industry, in making its great advance, has had, as I have said, to face considerable handicaps. I refer, I need scarcely say, to the lack of facilities in this country for higher technological education. It is true that in six or seven of our universities we have departments of chemical engineering and I have not a word to say against their efficiency and merit. But this is not nearly enough for a great country which depends to such a vital extent upon paying for its food and raw materials by exporting manufactures of various sorts and kinds. Chemical engineering is an exceptionally difficult branch of technology. It lies at the root of so many forms of industry. In the field of industrial atomic power it is the efficiency of the chemical recovery which decides whether or not the plant is an economic proposition. . . .

It is no use, in this connexion, refusing to face one of the obstacles which stands in the way, namely the ridiculous intellectual snobbery concerning technology which unhappily pervades this country. For some obscure reason it is considered in many influential circles that technological competence is not really on a par socially or intellectually with a knowledge of the older subjects. It would be really amusing (if it were not so tragic) to see how arts men, whose knowledge of the rudiments of technology is not even up to the standard of '1066 and all that', have the impudence to look down upon people who know far more about the arts subjects than the arts men do of technology. They seem to consider it quite natural and normal not to know how soda is made or how electricity is produced provided they once learnt something – which they have usually forgotten – about the mistresses of Charles II or the divagations of Alcibiades.

Quite frankly I resent this attitude very much. I think it more important to know about the properties of chlorine than about the improprieties of Clodius; or about the behaviour of crystals than about the misbehaviour of Christina. Surely it is more important to know what a calorie is than what Caligula did; and anyhow what catalysts do is certainly more useful and less objectionable than what Catiline did.

The people who are keeping this country alive are the people who are producing saleable products whether it be in the chemical, the engineering or the textile fields. It may be very interesting and amusing to read about the amours of French kings, but it is the people who read about and know about, and sometimes themselves create, new technological processes who enable the country to live. It takes just as much brains to become a first-class technologist as to be a first-class lawyer – probably more. I consider it scandalous that this is not recognized by the country as a whole.

I have fought this corner for the last 20 years, unhappily not with any great success. The forces of inertia, or perhaps it would be technically more correct to say the frictional forces, have been too much for me. Even official Government pronouncements in our favour have not been enough. The men who run our administrations are scarcely ever technologically trained and very seldom technologically minded and, despite what is said, somehow or other nothing very much gets done. . . .

Winston Churchill

Part of 'Fifty years hence', a chapter of *Thoughts and Adventures*, 1932

There is no doubt that this evolution will continue at an increasing rate. We know enough to be sure that the scientific achievements of the next fifty years will be far greater, more rapid and more surprising, than those we have already experienced. The slide-lathe enabled machines of precision to be made, and the power of steam rushed out upon the world. And through the steam-clouds flashed the dazzling lightning of electricity. But this is only a beginning. High authorities tell us that new sources of power, vastly more important than any we yet know, will surely be discovered. Nuclear energy is incomparably greater than the molecular energy which we use to-day.

The coal a man can get in a day can easily do five hundred times as much work as the man himself. Nuclear energy is at least one million times more powerful still. If the hydrogen atoms in a pound of water could be prevailed upon to combine together and form helium, they would suffice to drive a thousand horse-power engine for a whole year. If the electrons – those tiny planets of the atomic system – were induced to combine with the nuclei in the hydrogen the horsepower liberated would be 120 times greater still. There is no question among scientists that this gigantic source of energy exists. What is lacking is the match to set the bonfire alight, or it may be the detonator to cause the dynamite to explode. The Scientists are looking for this.   .

The discovery and control of such sources of power would cause changes in human affairs incomparably greater than those produced by the steam-engine four generations ago. Schemes of cosmic magnitude would become feasible. Geography and climate would obey our orders. Fifty thousand tons of water, the amount displaced by the *Berengaria*, would, if exploited as described, suffice to shift Ireland to the middle of the Atlantic. The amount of rain falling yearly upon the Epsom race-course would be enough to thaw all the ice at the Arctic and Antarctic poles. The changing of one element into another by means of temperatures and pressures would be far beyond our present reach, would transform beyond all description our standards of values. Materials thirty times stronger than the best steel would create engines fit to bridle the new forms of power. Communications and transport by land, water and air would take unimaginable forms, if, as is in principle possible, we could make an engine of 600 horsepower weighing 20 lb. and carrying fuel for a thousand hours in a tank the size of a fountain-pen. Wireless telephones and television, following naturally upon their present path of development, would enable their owner to connect up with any room similarly installed, and hear and take part in the conversation as well as if he put his head in through the window. The congregation of men in cities would become superfluous. It would rarely be necessary to call in person on any but the most intimate friends, but if so, excessively rapid means of communication would be at hand. There would be no more object in living in the same city with one's neighbour than there is to-day in living with him in the same house. The cities and the countryside would become indistinguishable. Every home would have its garden and its glade.

Up till recent times the production of food has been the prime struggle of man. That war is won. There is no doubt that the civilized races can produce or procure all the food they require. Indeed some of the problems which vex us to-day are due to the production of wheat by white men having exceeded their own needs, before yellow men, brown men and black men have learnt to demand and become able to purchase a diet superior to rice. But food is at present obtained almost entirely from the energy of the sunlight. The radiation from the sun produces from the carbonic acid in the air more or less complicated carbon compounds which give us our plants and vegetables. We use the latent chemical energy of these to keep our bodies warm, we convert it into muscular effort. We employ it in the complicated processes of digestion to repair and replace the wasted cells of our bodies. Many people of course prefer food in what the vegetarians call 'the second-hand form', i.e. after it has been digested and converted into meat for us by domestic animals kept for this purpose. In all these processes however ninety-nine parts of the solar energy are wasted for every part used.

Even without the new sources of power great improvements are probable here. Microbes which at present convert the nitrogen of the air into the proteins by which animals live, will be fostered and made to work under controlled conditions, just as yeast is now. New strains of microbes will be developed and made to do a great deal of our chemistry for us. With a greater knowledge of what are called hormones, i.e. the chemical messengers in our blood, it will be possible to control growth. We shall escape the absurdity of growing a whole chicken in order to eat the breast or wing, by growing these parts separately under a suitable medium. Synthetic food will, of course, also be used in the future. Nor need the pleasures of the table be banished. That gloomy Utopia of tabloid meals need never be invaded. The new foods will from the outset be practically indistinguishable from the natural products, and any changes will be so gradual as to escape observation.

If the gigantic new sources of power become available, food will be produced without recourse to sunlight. Vast cellars in which artificial radiation is generated may replace the cornfields of potato-patches of the world. Parks and gardens will cover our pastures and ploughed fields. When the times comes there will be plenty of room for the cities to spread themselves again. . . .

Paul Colinvaux

Chapter 3 of *Why Big Fierce Animals are Rare: An Ecologist's Perspective*, 1978

Animals come in different sizes, and the little ones are much more common than the big.

A typical small patch of woodland in any of the temperate lands of the North will contain hosts of insects and then nothing larger running about until we get to the size of small birds, which are much less numerous. Another size jump brings us to foxes, hawks, and owls, of which there may be only one or two. A fox is ten times the size of a song bird, which is ten times the size of an insect. If the insect is one of the predacious ground beetles of the forest floor, which hunt among the leaves like the wolf-spiders, then it, in turn, is ten times bigger than the mites and other tiny things that they both hunt.

The animals in this system of living do indeed come in very distinct sizes. There are, of course, some in-between ones, but not many. Squirrels in the upper size range seem obvious, but I am hard put to find something between an insect and a small bird unless it is a newt or lizard, neither of them very prominent denizens of a temperate woodland. Slugs and snails are toward the size of caterpillars. Shrews and toads are near the size of song birds. Even a snake can be thought of as an odd-shaped hawk.

In the wood as elsewhere there are distinctly different sizes, and the little ones are the most common. The same sort of thing exists in the sea in even odder form, for in the open sea the really tiny things are plants; the microscopic diatoms and other algae. Ten times bigger than these (give or take a few times) are the animals of the plankton, the copepods and the like. Bigger still are the shrimps and fish that hunt those copepods. Then another jump brings us to herrings, then to sharks, or killer whales. In any one place in the sea, this clumping of life into different sizes is the normal thing.

In the sea the rarity of the large is also most clearly shown. Great white sharks are extremely rare, and the other kinds of shark are scattered pretty thinly over the seas too. Fish of the herring size are vastly more common than sharks, but, even so, the number that are seen in a casual dive in the sea is seldom immense. If you drift, and focus your eyes just outside the facemask, however, myriad darting specks of the smaller animals may become visible. If you later take some of that same water and spin it in a centrifuge, there is likely to be a thin green scum in the bottom made up of an almost uncountable multitude of independent, tiny plants.

The tiny things of woodland and sea are immensely common; bigger things are a whole jump bigger and a whole jump less common; and so on until we reach the largest and rarest animals of all. A like pattern can be found in tropical forests, Irish bogs, or just about anywhere else. It is an extraordinary thing but true that life comes in size-fractions which, for all the blending and exceptions that can be found by careful scrutiny, are remarkably distinct. Animals in the larger sizes are comparatively rare.

Charles Elton of Oxford pointed out this strange reality half a century ago. Elton went adventuring on Spitzbergen, an arctic island covered with treeless tundra, where the animals move about in the open and where particularly he could follow an arctic fox as it went about its daily affairs. Arctic foxes can be delightfully tame. On St George Island in the Bering Sea one tried to take sandwiches from my pocket as I sat upon a rock. Elton followed his foxes and pondered their activities through a summer that was to be one of the most important an ecologist ever spent.

The foxes caught the summer birds of the tundra – the ptarmigan, sandpipers, and buntings; and these birds were at once a size-jump smaller than the foxes and much more numerous. The ptarmigan ate the fruit and leaves of tundra plants, but the sandpipers and buntings ate insects and worms, which were again a size-jump smaller as well as being more numerous. The foxes also ate seagulls and eider ducks, smaller and more numerous than the foxes, and these birds ate the tiny abundant life of the sea. Elton not only saw all this but, as Sherlock Holmes often lectured Watson, he *observed* it also. That small things were common and large things rare has been known by everybody since the dawn of thought, but Elton pondered it as Newton once pondered a falling apple, and he knew he was watching something odd. Why should large animals be so remarkably rare? And why should life come in discrete sizes?

Elton's summer on Spitzbergen gave him the answer to the second of these questions even as he posed it. The discrete sizes came about from the mechanics of eating and being eaten. He had seen a fox eat a sandpiper and a sandpiper eat a worm. These animals of different sizes were linked together by invisible chains of eating and being eaten. Foxes had to be big enough, and active enough, to catch and eat the birds on which they preyed; and the birds likewise must overpower, and engulf at a single swallow, the animals on which they fed. The normal lot of an animal was to be big enough to vanquish its living food with ease, and usually to be able to stuff it down its throat whole or nearly so. As one moves from link to link of a food chain the animals got roughly ten times bigger. Life comes in discrete sizes because each kind must evolve to be much bigger than the thing it eats.

Elton's conclusions were obviously true in a very general way. The communities of woodland and ocean on which his thinking was based seemed to conform very nicely. Life in those communities did come in different sizes, and it seemed that the sizes had grown discrete because each kind had evolved to be much bigger than the thing it ate. But many exceptions to the general principle of food size come to mind: wolves, lions, internal parasites, elephants, and baleen whales. There are many animals that are either smaller than their food, such as wolves or parasites, or else absurdly bigger like whales. But a closer look at any of these animals shows them to be instructive exceptions, if true exceptions to the rule all the same.

Land herbivores do not fit the Eltonian model, at least not completely, because land plants provide different-sized mouthfuls for different sizes of animals. You do not have to kill an entire land plant in order to eat it, you just tear off a suitable piece, a shoot, some grass blades, a berry, a bite out of a leaf. Food chains based on vegetation could start with many different sizes of plant-eating animal because squirrels, caterpillars, and elephants share the same food. Even so there does not seem to be a complete continuum of sizes amongst vegetarians, at least in any one place. Both big and little plant-eaters exist in a forest or a prairie and there is not much difficulty in sorting them into sizes. This is because the predators of plant-eaters do have to be size conscious when they look for food. A selection pressure acts downward along the food-chains, as herbivores evolve sizes that let them escape even as carnivores evolve sizes that enable them to catch skilfully. It is as important to be of a size that does not fit in someone else's mouth as it is to have a mouth suited to the size of one's own prey. So natural selection tends to preserve size classes even when food chains start with a pablum of meadow-forage or forest.

Wolves sometimes obey Eltonian principles, as when they hunt singly for rodents and small game, but they have evolved the trick of packing-up to haul down bigger prey, in winter, when they are freed from family cares and can go out in gangs. Other pack-hunting animals work variants on this method. And all large carnivores have had their sizes adjusted to the needs of killing, rather than of engulfing, so that a lion needs to be big enough to pull down an ailing zebra, but no bigger.

Parasites are smaller than their food, for obvious reasons, but their activities still tend to separate the animals on parasite food chains into different sizes with every link as was described before the coming of ecology in a jingle by Jonathan Swift:

> Big fleas have little fleas
> Upon their backs to bite 'em
> And little fleas have lesser fleas
> And so *ad infinitum*.

Very special sea animals such as whales are even more instructive, and we discuss them at the end of this chapter. But otherwise in the sea the pattern of size tends to conform very well to the simplest interpretation of the workings of food size. This is because the sea plants are tiny, individual, and have to be hunted and killed by those who would live off them (seaweeds of the coasts are of trivial importance in the wide oceans). So in the sea a rather complete set of steps runs up the food chains from the smallest plants, through crustaceans and fish, to great white sharks.

Thinking these Eltonian thoughts brings up another of nature's conundrums, 'Why are land plants big but sea plants small?' But that must wait for another chapter.

Now there was the matter of rarity. Elton showed that there had to be size-jumps as one went up food chains, and that the animals on the upper end had to be big. But why should the big be so rare? And very rare they are. One has only to compare the number of sharks to the number of herrings, or warblers to caterpillars, to see this. With every jump in size an even mightier loss occurs in numbers. Elton coined a term to describe this fact of life; he called it 'The Pyramid of Numbers'. He saw in his mind's eye a mighty host of tiny animals supporting on their backs a much smaller army of animals ten times as big. And this array supported, in turn, other animals ten times bigger still, but these were a select few. It was a graph of life he imagined with numbers of individuals along the horizontal axis, and position in the chain of eating, together with size, on the vertical. His vision saw the functioning of animal communities like the profile of Zoser's step pyramid at Saqqara, a triangular edifice built of stacked square-ended layers so that the summit could be reached by four or five giant steps. When ecologists forgather they call this result the 'Eltonian Pyramid'. Now, why should there be pyramids of numbers in nature wherever we look, from the arctic tundras to tropical forests and the open spaces of the sea? Why should large animals, particularly large hunting animals, always be so amazingly rare?

It is tempting to say that no problem exists, that it stands to reason that there cannot be as many big things as little. But this claim suggests that the Eltonian pyramid reflects no more than the elementary facts of spatial geometry. There is clearly no shortage of actual space to hold more big animals. On Spitzbergen, for instance, each fox had acres and acres to run around in, and the world oceans could hold mind-boggling quantities of the large sharks and killer whales who are the top carnivores of the sea. Large plants are crammed together on the earth in astounding numbers so that we call the result a 'forest'. Only the large animals are discriminated against.

A second tempting argument is to say that there is a finite amount of flesh (what ecologists call 'biomass') to go round and that this chunk of flesh could be used either to make a few big bodies or to make very many little ones. The big are rare because they take large slices from their cake. This assertion is true as far as it goes, but it does not go nearly far enough. If, instead of counting the animals in the different size levels of the pyramid, one weighed them, one finds that there is vastly more flesh in the smaller classes, a greater standing crop of life as well as more numerous individuals. All the insects in a woodlot weigh many times as much as all the birds; and all the songbirds, squirrels, and mice combined weigh vastly more than all the foxes, hawks, and owls combined. The pyramid of numbers is also a pyramid of mass, and the problem remains unsolved. Why is there so little living tissue in the larger animal sizes?

Elton did not have the answer. He thought it might be because little animals reproduced very quickly (true, they do – compare the egg output of butterflies with that of the birds that eat caterpillars) and that rapid reproduction was the key to vast populations. But this is to fall into the age-old error of biologists and theologians alike, the error that says numbers are set by breeding strategy. We have discussed this non-Darwinian idea in the last chapter. Numbers are set by the opportunities for one's way of life, not by the way one breeds. Professorships set the limit to the population of professors, not the productive output of graduate schools. The fact that large animals are rare cannot have anything to do with their reproductive drives. Elton's explanation will not do.

It took nearly twenty years for the corporate body of science to come up with the answer to the question Elton posed in 1927. Raymond Lindeman and Evelyn Hutchinson did so at Yale by thinking of food and bodies as calories rather than as flesh.

A unit of biomass or flesh represents a unit of potential energy that is measured in calories. If we burn a chunk of protein we liberate so many calories of heat, and if we burn a chunk of fat we get more calories still. This is now common knowledge to the affluent peoples of the West who worry about the calories in their food lest they become obese. In the 1930s and 1940s even illiterate Hollywood starlets knew this, but biologists wakened to the idea of the calorie rather more slowly. Yet in the use of food as calories lay the answer to the rarity of the large and fierce.

Measuring an animal's flesh in calories also alerts one's mind to the vital fact that bodies represent fuel as well as vessels for the soul. An animal continually burns up its fuel supply to do the work of living, puffing the exhaust gases out of the smokestacks of its mouth and nostrils and sending the calories off to outer space as radiant heat. The animal uses up its flesh, replacing the lost substance by eating more food, then burning most of this up too. This process of consuming matter by the fires of life goes on in every level of the Eltonian pyramids, and the fires are continually fresh-stoked by the plants on which the animal pyramids rest. At each successive level in the pyramids, the animals have to make do with the fuel (food) that can be extorted from the level below. But they can only extort some fraction of what the level below had not itself used up, and with this tithe the denizens of the upper layers must both make their own bodies and fuel their lives. Which is why their numbers are only a fraction of the numbers below, which is to say why they are rare.

The ultimate furnace of life is the sun, streaming down calories of heat with never-fainting ray. On every usable scrap of the earth's surface a plant is staked out to catch the light, its green array of energy receptors and transducers tuned and directed to the glowing source like the gold-plated cells on the arms of a satellite. In those green transducers we call leaves, the plants synthesise fuel, taking a constant allotment of the streaming energy of the sun. Some of this fuel they use to build their bodies, but some they burn to do the work of living. Animals eat those plants, but they do not get all the plant tissue, as we know because the earth is carpeted brown with rotting debris that has not been part of an animal's dinner. Nor can the animals ever get the fuel the plants have already burned. So there cannot be as much animal flesh on the earth as there is plant flesh. It is possible for large plants to be vastly abundant and ranked side by side, but animals of the same size would have to be thinly spread out because they can only be a tenth as abundant.

This would be true even if all animals were vegetarian. But they are not. For flesh eaters, the largest possible supply of food calories they can obtain is a fraction of the bodies of their plant-eating prey, and they must use this fraction both to make bodies and as a fuel supply. Moreover their bodies must be the big active bodies that let them hunt for a living. If one is higher still on the food chain, an eater of a flesh-eater's flesh, one has yet a smaller fraction to support even bigger and fiercer bodies. Which is why large fierce animals are so astonishingly (or pleasingly) rare.

Thus was the grandest pattern of rarity and abundance in the world explained by two men at Yale in the 1940s. Ways of life were bumping against that most fundamental of physical restraints, the supply of energy.

As the realisation of what Lindeman and Hutchinson had done for natural history percolated through the consciousness of biology in the fifties and sixties a thrill of self-respect began to throb in its younger practitioners. Here the pattern of field experience was linked to the fundamental laws of physics. We were talking of energy degraded step by step as it flowed down food chains, losing its power to do work and pouring steadily away to the sink of heat. The grand pattern of life that Elton had seen on Spitzbergen and that countless naturalists had intuitively known before was clearly and directly a consequence of the second law of thermodynamics.

We can now understand why there are not fiercer dragons on the earth than there are; it is because the energy supply will not stretch to the support of super-dragons. Great white sharks or killer whales in the sea, and lions and tigers on the land, are apparently the most formidable animals the contemporary earth can support. Even these are very thinly spread. One may swim many lifetimes in the world oceans without encountering a great white shark; and an ancient Chinese proverb asserts that a hill shelters only one tiger. Evolutionary principle tells us that the existence of these animals creates a theoretical possibility for other animals to evolve to eat them, but the food calories to be won from the careers or niches of hunting great white sharks and tigers are too few to support a minimum population of animals as large and horribly ferocious as these would have to be. Such animals, therefore, have never evolved. Great white sharks and tigers represent the largest predators that the laws of physics allow the contemporary earth to support.

But here we run into what seems to be the first real difficulty of the argument. There are living animals that are much larger than tigers and sharks, and there have been some very big ones in the past. How does their existence square with our interpretation of the second law of thermodynamics?

Elephants and the big, cloven-hoofed animals are larger than tigers. In the past there have been even bigger mammals, such as giant ground-sloths and *Titanotherium*, a beast like an overgrown elephant and the largest mammal ever. Ther have also been the largest reptiles of the Mesozoic, the ponderous dinosaurs: *Stegosaurus*, *Brontosaurus*, *Iguanadon*. None of these animals poses any difficulty for the model. They have all been plant-eaters. In the strict Eltonian model the plant-eaters are small, and indeed in life most of them actually are. In the open sea this rule that plant-eaters must be small is strictly enforced because the drifting plants are so tiny that only very small animals can make a successful living by eating them. But on land, plants often appear as continuous mats of leaves, which we call vegetation, and it is possible for enormous sluggish animals to slurp them up without much nicety in the hunting. Masses of energy are available in the plant-eating niches at the bottom of the Eltonian pyramids, with the result that viable populations of even enormous animals can be supported. The brontosaur and the elephant alike, therefore, leave both our belief in the energy-flow model and the second law of thermodynamics intact.

That leaves two trickier kinds of animals to explain away: the great baleen whales of the contemporary oceans, which are the largest animals ever to have lived, and the flesh-eating dinosaurs such as *Tyranosaurus rex*. These are both meat-eating animals, and they are impressively bigger than great white sharks or tigers.

The baleen whales have learned to cheat, hunting their food in non-Eltonian ways. Essential to the normal structure of the Eltonian pyramid was that every carnivorous animal should have a direct relationship to the size of its food, being big enough to catch and eat it but not so big that the food item should prove a trivial mouthful not worth the effort of hunting. On this model, the food of a blue or white whale should be several feet long. But it is not. The whales cheat with their sieves of baleen, which let them strain from the surface of the sea the tiny shrimps called krill in huge numbers and with little effort. The whales have cut out the middlemen, avoiding all the energy losses that would have accrued if the krill had been passed to a fish and that fish passed to a bigger fish before the whale had its chance at it like any other Eltonian feeder. So the whales, although not plant eaters, feed very low on food chains where the energy supply is still comparatively large. Floating as they do in the sea, they use little energy in their sluggish hunting, paddling quietly along with their mouths open, straining the meat out of the oceanic soup. So the apparent exception of the whales is no exception at all, and our model may stand.

Tyranosaurus rex is more difficult for the argument. Tyranosaurs were huge carnivorous dinosaurs, often pictured as a great green kangaroo-like form with a hideous toad-like head, nightmare teeth, and a pair of useless little flapping arms dangling below the ugly neck. An animal answering to the name of *Tyranosaurus* of this size certainly existed, for we have specimens of all his bones. He was several times larger than lions or tigers, or indeed of any other recorded predator. What enabled it to escape the constraints apparently placed on all its successors by the second law of thermodynamics?

It is useful to note first that the tyranosaur fed at the same level as its modern successors, the big cats, and at the same level as the baleen whales in the sea. If fed on plant-eaters relatively low in the food chain, close to the bottom of the Eltonian pyramid, where there was still much energy to be won. A large body, therefore, does not seem hopelessly out of the question. We know that there were many kinds of very large herbivores about in the tyranosaur's time, animals that, in the absence of pack-hunting predators such as dogs, could be overcome only by very powerful attackers. So we might conclude that the necessity for Mesozoic predators to be large and ferociously active is self-evident. There was nothing else to get at the meat so massively on the hoof, so natural selection provided *Tyranosaurus rex*.

I have always been unhappy about this reasoning. If natural selection could fashion a tyranosaur at that time, why not in all subsequent time? Why in particular was there nothing like a tyranosaur in the great age of mammals, that later part of the Tertiary epoch when all the plainslands of the earth held herds of game that make the herds of modern Africa seem trivial by comparison? I have felt compelled to conclude that the constraints on the size of ferociously active predators that has been applied throughout the age of mammals ought to have applied to the reptiles of the Mesozoic era also. By thinking thus I manoeuvre myself into the position of saying that, on ecological grounds, the *Tyranosaurus rex* did not exist. And yet there the bones are, indubitably the ones of a large flesh-eating animal of the size claimed. It was with a sense of inward peace that I saw a drawing of a recent attempt to put the bones together differently.

The classic picture of the hopping, predacious tyrant-lizard is derived from nineteenth-century reconstructions of the animal. The new reconstruction, first published in *Nature* in 1968, shows the animal to be a waddling, slow-moving beast, not at all the sort one can imagine dashing after a herd of galloping brontosauri. But it probably got them all the same, picking out the sick and the dying, often getting them only as carrion. The tyranosaur was not a ferociously active predator. It did not stand upright, nor did it hop. It held that massive body horizontally, perhaps able to move swiftly for short periods as it balanced its motion with the long tail. But most of its days were spent lying on its belly, a prostration that conserved energy and from which it periodically roused itself, lifting its great bulk on those two little arms in front until it could balance on the thick walking legs. The tyranosaur did indeed support a large mass by meat-eating, but it escaped the energy-consuming price of being active in order to overcome prime specimens of the giant prey it ate. It managed on land essentially the same stratagem that the baleen whales managed in the sea; it found a non-Eltonian way of getting the meat of plant-eaters without having to hunt them properly. Nothing like it has been seen since because the true active predators of the age of mammals were able to clean up the meat supplies before a sluggish beast such as a tyranosaur could get to them. And active predators might even have eaten the tyranosaur itself.

Tyranosaurus rex, as popularly portrayed, is a myth. But it is probably safe to say that it will be as durable as any other myth in our culture. The size and ferocity of real-life predators is restricted to the scale of a tiger, and even these must always be rare. The second law of thermodynamics says so.

E. J. H. Corner

Extracts from *The Life of Plants*, 1964

There is a giant tree, pre-eminent in a forest that stretches to the skyline. On its canopy birds and butterflies sip nectar. On its branches orchids, aroids, and parasitic mistletoes offer flowers to other birds and insects. Among them ferns creep, lichens encrust, and centipedes and scorpions lurk. In the rubble that falls among the epiphytic roots and stems, ants build nests and even earthworms and snails find homes. There is a minute munching of caterpillars and the silent sucking of plant bugs. On any of these things, plant or animal, fungus may be growing. Through the branches spread spiders' webs. Frogs wait for insects, and a snake glides. There are nests of birds, bees, and wasps. Along a limb pass wary monkeys, a halting squirrel, or a bear in search of honey; the shadow of an eagle startles them. Through dead snags fungus and beetle have attacked the wood. There are fungus brackets nibbled round the edge and bored by other beetles. A woodpecker taps. In a hole a hornbill broods. Where the main branches diverge, a strangling fig finds grip, a bushy epiphyte has temporary root, and hidden sleeps a leopard. In deeper shade black termites have built earthy turrets and smothered the tips of a young creeper. Hanging from the limbs are cables of lianes which have hoisted themselves through the undergrowth and are suspended by their grapnels. On their swinging stems grows an epiphytic ginger whose red seeds a bird is pecking. Where rain trickles down the trunk filmy ferns, mosses, and slender green algae maintain their delicate lives. Round the base are fragments of bark and coils of old lianes, on which other ferns are growing. Between the buttress-roots a tortoise is eating toadstools. An elephant has rubbed the bark and, in its deepened footmarks, tadpoles, mosquito larvae, and threadworms swim. Pigs squeal and drum in search of fallen fruit, seeds, and truffles. In the humus and undersoil, insects, fungi, bacteria, and all sorts of 'animalculae' participate with the tree roots in decomposing everything that dies.

This tree is not alone. The forest consists not solely of its kind, not even of a few, but of hundreds, each with its specific size, shape, bark, leaf, flower, fruit, and wood attractive to its particular following of green plants, fungi, and animals. Thus comes the richness of these forests, most prolific in natural history, and thence sprung the superiority of the creatures that intelligently traverse them. . . .

Willow branches spread widely. Narrow leaves are set outwards on short stalks, roughly in two rows along slender twigs. The sapling grows erect. Its leaves are borne in several rows round the stem, and the branches that come from buds in their axils grow upwards in all directions. As they branch in turn, the leaf arrangement becomes simpler and the twigs tend also to be set in two rows. Thus by branching to occupy space, and then by restricting the branching to consolidate the position, the tree builds a scaffolding that supports and exposes to the light with least overlapping the peripheral canopy of foliage. At length, long branches sag to the ground.

The poplar is the nearest ally of the willow. Its leaves are differently shaped. They hang and clatter on long stalks set round the stouter twig, which grows upward. The twig can branch in all directions but, with better illumination on the outside of the crown, the outer buds develop more strongly and raise the crown outwards and upwards. Inner branches, if they are formed, become overshadowed and in a year or two will die off. The erect growth of the twigs repeats the habit of the sapling. The lofty poplar therefore is more generalized in its tree-form than the willow. It keeps the all-round or radial construction of upward growth, whereas the willow changes to a flattened system of leaves and branches, making the sprays of foliage that are thrust across the path of light. Willows may weep but poplars express themselves with the stiffness of the Lombardy. . . .

From all over the earth – tundra, desert, steppe, mountain, ocean, island, and lake – we gather the threads of plant life on land, and we trace them in the canopy of the forest, which first fitted plants for their life in the desert and on the mountain, down to the beginnings of trees flowering, fruiting, seeding, and even sporing on the river flats. Too little is known of these places, where the detritus of the land began to consolidate and its plants to root. There will be time enough to learn about the ocean, where the plant-cell evolved, and there will probably be time enough to learn about the seashore where this cell grew into the plant form. But the forests, which show how trees were made, are going. They are vanishing nowhere faster than from the alluvial plains where the vestiges of the last creative phase of plant life, that prepared the way for the modern world, may survive. The modern mouth is the people's, and theirs the new retaliation. Before machines the forest is defenceless. Human progress is clearing it with gathering speed to plant crops of quick returns. The botanist must hurry if he would take the opportunity that a few brief centuries of his science have revealed; for soon there will be rice-field to every river-brink. The unmindful tree begot, indeed, the seed of its own destruction.

George W. Corner

Extracts from *The Hormones in Human Reproduction*, 1943

I do not pretend to write this chapter in cool detachment. Its theme-word *progesterone* has for me connotations that will never be found in the dictionary. In the first place I invented the word myself, as far at least as the letter 't', as will be explained hereafter. In the second place, it recalls memories of bafflement, comedy, hard work, and modest success. Can I forget the time I went racing up the steps of the laboratory in Rochester, carrying a glass syringe that contained the world's entire supply of crude progesterone, stumbled and fell and lost it all? Or the day Willard Allen showed me his first glittering crystals of the hormone, chemically pure at last? . . .

The collection of evidence begins with a scene poignant enough, indeed, for a novel. In 1900 the great embryologist of Breslau, Gustav Born, lay dying. Scientist to the last, his mind was full of a hypothesis he knew he could not live to test and which he could not bear to leave untried. To his bedside, therefore, he summoned one of his former students, the rising young gynecologist Ludwig Fraenkel. To him Born imparted his thought that the corpus luteum is indeed an organ of internal secretion, and moreover that its function must be concerned with the protection of the early embryo. This guess about its specific function, like Prenant's about its general nature, was brilliant and novel in its day, even though to us in retrospect when we consider that the corpus luteum is present only when the egg is available for development, such a function seems probable indeed. So it seemed then to Fraenkel, whose task it was to devise the experiments by which Born's conjecture could be put to rigorous test. . . .

Charles Darwin

An extract from *Journal of Researches*, 1890

When we arrived at the head of the lagoon, we crossed a narrow islet, and found a great surf breaking on the windward coast. I can hardly explain the reason, but there is to my mind much grandeur in the view of the outer shores of these lagoon-islands. There is a simplicity in the barrier-like beach, the margin of green bushes and tall cocoa-nuts, the solid flat of dead coral-rock, strewed here and there with great loose fragments, and the line of furious breakers, all rounding towards either hand. The ocean throwing its waters over the broad reef appears an invincible, all-powerful enemy; yet we see it resisted, and even conquered, by means which at first seem most weak and inefficient. It is not that the ocean spares the rock of coral; the great fragments scattered over the reef, and heaped on the beach, whence the tall cocoa-nut springs, plainly bespeak the unrelenting power of the waves. Nor are any periods of repose granted. The long swell caused by the gentle but steady action of the trade wind, always blowing in one direction over a wide area, causes breakers, almost equalling in force those during a gale of wind in the temperate regions, and which never ceased to rage. It is impossible to behold these waves without feeling a conviction that an island, though built of the hardest rock, let it be porphory, granite, or quartz, would ultimately yield and be demolished by such an irresistible power. Yet these low, insignificant coral-islets stand and are victorious: for here another power, as an antagonist, takes part in the contest. The organic forces separate the atoms of carbonate of lime, one by one, from the foaming breakers and unite them into a symmetrical structure. Let the hurricane tear up its thousand huge fragments; yet what will that tell against the accumulated labour of myriads of architects at work night and day, month after month? Thus do we see the soft and gelatinous body of a polypus, through the agency of the vital laws, conquering the great mechanical power of the waves of an ocean which neither the art of man nor the inanimate works of nature could successfully resist. . . .

An extract from his *Autobiography*, 1876

After my return to England it appeared to me that by following the example of Lyell in Geology, and by collecting all facts which bore in any way on the variation of animals and plants under domestication and nature, some light might perhaps be thrown on the whole subject. My first note-book was opened in July 1837. I worked on true Baconian principles, and without any theory collected facts on a wholesale scale, more especially with respect to domesticated productions, by printed enquiries, by conversation with skilful breeders and gardeners, and by extensive reading. When I see the list of books of all kinds which I read and abstracted, including whole series of Journals and Transactions, I am surprised at my industry. I soon perceived that selection was the keystone of man's success in making useful races of animals and plants. But how selection could be applied to organisms living in a state of nature remained for some time a mystery to me.

In October 1838, that is, fifteen months after I had begun my systematic enquiry, I happened to read for amusement 'Malthus on Population', and being well prepared to appreciate the struggle for existence which everywhere goes on from long-continued observation of the habits of animals and plants, it at once struck me that under these circumstances favourable variations would tend to be preserved, and unfavourable ones to be destroyed. The result of this would be the formation of new species. Here then I had at last got a theory by which to work; but I was so anxious to avoid prejudice, that I determined not for some time to write even the briefest sketch of it. In June 1842 I first allowed myself the satisfaction of writing a very brief abstract of my theory in pencil in 35 pages; and this was enlarged during the summer of 1844 into one of 230 pages, which I had fairly copied out and still possess.

But at that time I overlooked one problem of great importance; and it is astonishing to me, except on the principle of Columbus and his egg, how I could have overlooked it and its solution. This problem is the tendency in organic beings descended from the same stock to diverge in character as they become modified. That they have diverged greatly is obvious from the manner in which species of all kinds can be classed under genera, genera under families, families under sub-orders and so forth; and I can remember the very spot in the road, whilst in my carriage, when to my joy the solution occurred to me; and this was long after I had come to Down. The solution, as I believe, is that the modified offspring of all dominant and increasing forms tend to become adapted to many and highly diversified places in the economy of nature.

Early in 1856 Lyell advised me to write out my views pretty fully, and I began at once to do so on a scale three or four times as extensive as that which was afterwards followed in my *Origin of Species*; yet it was only an abstract of the materials which I had collected, and I got through about half the work on this scale. But my plans were overthrown, for early in the summer of 1858 Mr. Wallace, who was then in the Malay archipelago, sent me an essay 'On the Tendency of Varieties to depart indefinitely from the Original Type'; and this essay contained exactly the same theory as mine. Mr. Wallace expressed the wish that if I thought well of his essay, I should send it to Lyell for perusal.

The circumstances under which I consented at the request of Lyell and Hooker to allow of an abstract from my MS., together with a letter to Asa Gray, dated September 5, 1857, to be published at the same time with Wallace's Essay, are given in the *Journal of the Proceedings of the Linnean Society*, 1858, p. 45. I was at first very unwilling to consent, as I thought Mr. Wallace might consider my doing so unjustifiable, for I did not then know how generous and noble was his disposition. The extract from my MS. and the letter to Asa Gray had neither been intended for publication, and were badly written. Mr. Wallace's essay, on the other hand, was admirably expressed and quite clear. Nevertheless, our joint productions excited very little attention, and the only published notice of them which I can remember was by Professor Haughton of Dublin, whose verdict was that all that was new in them was false, and what was true was old. This shows how necessary it is that any new view should be explained at considerable length in order to arouse public attention.

In September 1858 I set to work by the strong advice of Lyell and Hooker to prepare a volume on the transmutation of species, but was often interrupted by ill-health, and short visits to Dr. Lane's delightful hydropathic establishment at Moor Park. I abstracted the MS. begun on a much larger scale in 1856, and completed the volume on the same reduced scale. It cost me thirteen months and ten days' hard labour. It was published under the title of the *Origin of Species*, in November 1859. Though considerably added to and corrected in the later editions, it has remained substantially the same book.

It is no doubt the chief work of my life. It was from the first highly successful. The first small edition of 1250 copies was sold on the day of publication, and a second edition of 3000 copies soon afterwards. Sixteen thousand copies have now (1876) been sold in England; and considering how stiff a book it is, this is a large scale. It has been translated into almost every European tongue, even into such languages as Spanish, Bohemian, Polish, and Russian. . . .

The success of the *Origin* may, I think, be attributed in large part to my having long before written two condensed sketches, and to my having finally abstracted a much larger manuscript, which was itself an abstract. By this means I was enabled to select the more striking facts and conclusions. I had, also during many years followed a golden rule, namely, that whenever a published fact, a new observation or thought came across me, which was opposed to my general results, to make a memorandum of it without fail and at once; for I had found by experience that such facts and thoughts were far more apt to escape from the memory than favourable ones. Owing to this habit, very few objections were raised against my views which I had not at least noticed and attempted to answer.

It has sometimes been said that the success of the *Origin* proved 'that the subject was in the air', or 'that men's minds were prepared for it'. I do not think that this is strictly true, for I occasionally sounded not a few naturalists, and never happened to come across a single one who seemed to doubt about the permanence of species. Even Lyell and Hooker, though they would listen with interest to me, never seemed to agree. I tried once or twice to explain to able men what I meant by Natural Selection, but signally failed. What I believe was strictly true is that innumerable well-observed facts were stored in the minds of naturalists ready to take their proper places as soon as any theory which would receive them was sufficiently explained. Another element in the success of the book was its moderate size; and this I owe to the appearance of Mr. Wallace's essay; had I published on the scale in which I began to write in 1856, the book would have been four or five times as large as the *Origin*, and very few would have had the patience to read it.

I gained much by my delay in publishing from about 1839, when the theory was clearly conceived, to 1859; and I lost nothing by it, for I cared very little whether men attributed most originality to me or Wallace; and his essay no doubt aided in the reception of the theory. I was forestalled in only one important point, which my vanity has always made me regret, namely, the explanation by means of the Glacial period of the presence of the same species of plants and of some few animals on distant mountain summits and in the arctic regions. This view pleased me so much that I wrote it out in extenso, and I believe that it was read by Hooker some years before E. Forbes published his celebrated memoir on the subject. In the very few points in which we differed, I still think that I was in the right. I have never, of course, alluded in print to my having independently worked out this view.

Hardly any point gave me so much satisfaction when I was at work on the *Origin*, as the explanation of the wide difference in many classes between the embryo and the adult animal, and of the close resemblance of the embryos within the same class. No notice of this point was taken, as far as I remember, in the early reviews of the *Origin*, and I recollect expressing my surprise on this head in a letter to Asa Gray. Within late years several reviewers have given the whole credit to Fritz Müller and Häckel, who undoubtedly have worked it out much more fully, and in some respects more correctly than I did. I had materials for a whole chapter on the subject, and I ought to have made the discussion longer; for it is clear that I failed to impress my readers; and he who succeeds in doing so deserves, in my opinion, all the credit.

This leads me to remark that I have almost always been treated honestly by my reviewers, passing over those without scientific knowledge as not worthy of notice. My views have often been grossly misrepresented, bitterly opposed and ridiculed, but this has been generally done, as I believe, in good faith. On the whole I do not doubt that my works have been over and over again greatly overpraised. I rejoice that I have avoided controversies, and this I owe to Lyell, who many years ago, in reference to my geological works, strongly advised me never to get entangled in a controversy, as it rarely did any good and caused a miserable loss of time and temper.

Whenever I have found out that I have blundered, or that my work has been imperfect, and when I have been contemptuously criticised, and even when I have been overpraised, so that I have felt mortified, it has been my greatest comfort to say hundreds of times to myself that 'I have worked as hard and as well as I could, and no man can do more than this.' I remember when in Good Success Bay, in Tierra del Fuego, thinking (and, I believe, that I wrote home to the effect) that I could not employ my life better than in adding a little to Natural Science. This I have done to the best of my abilities, and critics may say what they like, but they cannot destroy this conviction. . . .

Erasmus Darwin

Lines from *The Temple of Nature*, 1802

> ORGANIC LIFE beneath the shoreless waves
> Was born, and nurs'd in Ocean's pearly caves;
> First forms minute, unseen by spheric glass,
> Move on the mud, or pierce the watery mass;
> These, as successive generations bloom,
> New powers acquire, and larger limbs assume;
> Whence countless groups of vegetation spring,
> And breathing realms of fin, and feet, and wing. . . .

Lines from *The Botanic Garden*, 1791

> Soon shall thy arm, UNCONQUER'D STEAM! afar
> Drag the slow barge, or drive the rapid car;
> Or on wide-waving wings expanded bear
> The flying-chariot through the fields of air.
> —Fair crews triumphant, leaning from above,
> Shall wave their fluttering kerchiefs as they move;
> Or warrior-bands alarm the gaping crowd,
> And armies shrink beneath the shadowy cloud. . . .

Richard Dawkins

'The replicators', chapter 2 of *The Selfish Gene*, 1976

In the beginning was simplicity. It is difficult enough explaining how even a simple universe began. I take it as agreed that it would be even harder to explain the sudden springing up, fully armed, of complex order – life, or a being capable of creating life. Darwin's theory of evolution by natural selection is satisfying because it shows us a way in which simplicity could change into complexity, how unordered atoms could group themselves into ever more complex patterns until they ended up manufacturing people. Darwin provides a solution, the only feasible one so far suggested, to the deep problem of our existence. I will try to explain the great theory in a more general way than is customary, beginning with the time before evolution itself began.

Darwin's 'survival of the fittest' is really a special case of a more general law of *survival of the stable*. The universe is populated by stable things. A stable thing is a collection of atoms which is permanent enough or common enough to deserve a name. It may be a unique collection of atoms, such as the Matterhorn, which lasts long enough to be worth naming. Or it may be a *class* of entities, such as rain drops, which come into existence at a sufficiently high rate to deserve a collective name, even if any one of them is short-lived. The things which we see around us, and which we think of as needing explanation – rocks, galaxies, ocean waves – are all, to a greater or lesser extent, stable patterns of atoms. Soap bubbles tend to be spherical because this is a stable configuration for thin films filled with gas. In a space-craft, water is also stable in spherical globules, but on earth, where there is gravity, the stable surface for standing water is flat and horizontal. Salt crystals tend to be cubes because this is a stable way of packing sodium and chloride ions together. In the sun the simplest atoms of all, hydrogen atoms, are fusing to form helium atoms, because in the conditions which prevail there the helium configuration is more stable. Other even more complex atoms are being formed in stars all over the universe, and were formed in the 'big bang' which, according to the prevailing theory, initiated the universe. This is originally where the elements on our world came from.

Sometimes when atoms meet they link up together in chemical reaction to form molecules, which may be more or less stable. Such molecules can be very large. A crystal such as a diamond can be regarded as a single molecule, a proverbially stable one in this case, but also a very simple one since its internal atomic structure is endlessly repeated. In modern living organisms there are other large molecules which are highly complex, and their complexity shows itself on several levels. The haemoglobin of our blood is a typical protein molecule. It is built up from chains of smaller molecules, amino acids, each containing a few dozen atoms arranged in a precise pattern. In the haemoglobin molecule there are 574 amino acid molecules. These are arranged in four chains, which twist around each other to form a globular three-dimensional structure of bewildering complexity. A model of a haemoglobin molecule looks rather like a dense thornbush. But unlike a real thornbush it is not a haphazard approximate pattern but a definite invariant structure, identically repeated, with not a twig nor a twist out of place, over six thousand million million million times in an average human body. The precise thornbush shape of a protein molecule such as haemoglobin is stable in the sense that two chains consisting of the same sequences of amino acids will tend, like two springs, to come to rest in exactly the same three-dimensional coiled pattern. Haemoglobin thornbushes are springing into their 'preferred' shape in your body at a rate of about four hundred million million per second, and others are being destroyed at the same rate.

Haemoglobin is a modern molecule, used to illustrate the principle that atoms tend to fall into stable patterns. The point that is relevant here is that, before the coming of life on earth, some rudimentary evolution of molecules could have occurred by ordinary processes of physics and chemistry. There is no need to think of design or purpose or directedness. If a group of atoms in the presence of energy falls into a stable pattern it will tend to stay that way. The earliest form of natural selection was simply a selection of stable forms and a rejection of unstable ones. There is no mystery about this. It had to happen by definition.

From this, of course, it does not follow that you can explain the existence of entities as complex as man by exactly the same principles on their own. It is no good taking the right number of atoms and shaking them together with some external energy till they happen to fall into the right pattern, and out drops Adam! You may make a molecule consisting of a few dozen atoms like that, but a man consists of over a thousand million million million million atoms. To try to make a man, you would have to work at your biochemical cocktail-shaker for a period so long that the entire age of the universe would seem like an eye-blink, and even then you would not succeed. This is where Darwin's theory, in its most general form, comes to the rescue. Darwin's theory takes over from where the story of the slow building up of molecules leaves off.

The account of the origin of life which I shall give is necessarily speculative; by definition, nobody was around to see what happened. There are a number of rival theories, but they all have certain features in common. The simplified account I shall give is probably not too far from the truth.

We do not know what chemical raw materials were abundant on earth before the coming of life, but among the plausible possibilities are water, carbon dioxide, methane, and ammonia; all simple compounds known to be present on at least some of the other planets in our solar system. Chemists have tried to imitate the chemical conditions of the young earth. They have put these simple substances in a flask and supplied a source of energy such as ultraviolet light or electric sparks – artificial simulation of primordial lightning. After a few weeks of this, something interesting is usually found inside the flask: a weak brown soup containing a large number of molecules more complex than the ones originally put in. In particular, amino acids have been found – the building blocks of proteins, one of the two great classes of biological molecules. Before these experiments were done, naturally-occurring amino acids would have been thought of as diagnostic of the presence of life. If they had been detected on, say Mars, life on that planet would have seemed a near certainty. Now, however, their existence need imply only the presence of a few simple gases in the atmosphere and some volcanoes, sunlight, or thundery weather. More recently, laboratory simulations of the chemical conditions of earth before the coming of life have yielded organic substances called purines and pyrimidines. These are building blocks of the genetic molecule, DNA itself.

Processes analogous to these must have given rise to the 'primeval soup' which biologists and chemists believe constituted the seas some three to four thousand million years ago. The organic substances became locally concentrated, perhaps in drying scum round the shores, or in tiny suspended droplets. Under the further influence of energy such as ultraviolet light from the sun, they combined into larger molecules. Nowadays large organic molecules would not last long enough to be noticed: they would be quickly absorbed and broken down by bacteria or other living creatures. But bacteria and the rest of us are late-comers, and in those days large organic molecules could drift unmolested through the thickening broth.

At some point a particularly remarkable molecule was formed by accident. We will call it the *Replicator*. It may not necessarily have been the biggest or the most complex molecule around, but it had the extraordinary property of being able to create copies of itself. This may seem a very unlikely sort of accident to happen. So it was. It was exceedingly improbable. In the lifetime of a man, things which are that improbable can be treated for practical purposes as impossible. That is why you will never win a big prize on the football pools. But in our human estimates of what is probable and what is not, we are not used to dealing in hundreds of millions of years. If you filled in pools coupons every week for a hundred million years you would very likely win several jackpots.

Actually a molecule which makes copies of itself is not as difficult to imagine as it seems at first, and it only had to arise once. Think of the replicator as a mould or template. Imagine it as a large molecule consisting of a complex chain of various sorts of building block molecules. The small building blocks were abundantly available in the soup surrounding the replicator. Now suppose that each building block has an affinity for its own kind. Then whenever a building block from out in the soup lands up next to a part of the replicator for which it has an affinity, it will tend to stick there. The building blocks which attach themselves in this way will automatically be arranged in a sequence which mimics that of the replicator itself. It is easy then to think of them joining up to form a stable chain just as in the formation of the original replicator. This process could continue as a progressive stacking up, layer upon layer. This is how crystals are formed. On the other hand, the two chains might split apart, in which case we have two replicators, each of which can go on to make further copies.

A more complex possibility is that each building block has affinity not for its own kind, but reciprocally for one particular other kind. Then the replicator would act as a template not for an identical copy, but for a kind of 'negative', which would in its turn re-make an exact copy of the original positive. For our purposes it does not matter whether the original replication process was positive–negative or positive–positive, though it is worth remarking that the modern equivalents of the first replicator, the DNA molecules, use positive–negative replication. What does matter is that suddenly a new kind of 'stability' came into the world. Previously it is probable that no particular kind of complex molecule was very abundant in the soup, because each was dependent on building blocks happening to fall by luck into a particular stable configuration. As soon as the replicator was born it must have spread its copies rapidly throughout the seas, until the smaller building block molecules became a scarce resource, and other larger molecules were formed more and more rarely.

So we seem to arrive at a large population of identical replicas. But now we must mention an important property of any copying process: it is not perfect. Mistakes will happen. I hope there are no misprints in this book, but if you look carefully you may find one or two. They will probably not seriously distort the meaning of the sentences, because they will be 'first generation' errors. But imagine the days before printing, when books such as the Gospels were copied by hand. All scribes, however careful, are bound to make a few errors, and some are not above a little wilful 'improvement'. If they all copied from a single master original, meaning would not be greatly perverted. But let copies be made from other copies, which in their turn were made from other copies, and errors will start to become cumulative and serious. We tend to regard erratic copying as a bad thing, and in the case of human documents it is hard to think of examples where errors can be described as improvements. I suppose the scholars of the Septuagint could at least be said to have started something big when they mistranslated the Hebrew word for 'young woman' into the Greek word for 'virgin', coming up with the prophecy: 'Behold a virgin shall conceive and bear a son. . . .' Anyway, as we shall see, erratic copying in biological replicators can in a real sense give rise to improvement, and it was essential for the progressive evolution of life that some errors were made. We do not know how accurately the original replicator molecules made their copies. Their modern descendants, the DNA molecules, are astonishingly faithful compared with the most high-fidelity human copying process, but even they occasionally make mistakes, and it is ultimately these mistakes which make evolution possible. Probably the original replicators were far more erratic, but in any case we may be sure that mistakes were made, and these mistakes were cumulative.

As mis-copyings were made and propagated, the primeval soup became filled by a population not of identical replicas, but of several varieties of replicating molecules, all 'descended' from the same ancestor. Would some varieties have been more numerous than others? Almost certainly yes. Some varieties would have been inherently more stable than others. Certain molecules, once formed, would be less likely than others to break up again. These types would become relatively numerous in the soup, not only as a direct logical consequence of their 'longevity', but also because they would have a long time available for making copies of themselves. Replicators of high longevity would therefore tend to become more numerous and, other things being equal, there would have been an 'evolutionary trend' towards greater longevity in the population of molecules.

But other things were probably not equal, and another property of a replicator variety which must have had even more importance in spreading it through the population was speed of replication or 'fecundity'. If replicator molecules of type A make copies of themselves on average once a week, while those of type B make copies of themselves once an hour, it is not difficult to see that pretty soon type A molecules are going to be far outnumbered, even if they 'live' much longer than B molecules. There would therefore probably have been an 'evolutionary trend' towards higher 'fecundity' of molecules in the soup. A third characteristic of replicator molecules which would have been positively selected is accuracy of replication. If molecules of type X and type Y last the same length of time and replicate at the same rate, but X makes a mistake on average every tenth replication while Y makes a mistake only every hundredth replication, Y will obviously become more numerous. The X contingent in the population loses not only the errant 'children' themselves, but also all their descendants, actual or potential.

If you already know something about evolution, you may find something slightly paradoxical about the last point. Can we reconcile the idea that copying errors are an essential prerequisite for evolution to occur, with the statement that natural selection favours high copying-fidelity? The answer is that although evolution may seem, in some vague sense, a 'good thing', especially since we are the product of it, nothing actually 'wants' to evolve. Evolution is something that happens, willy-nilly, in spite of all the efforts of the replicators (and nowadays of the genes) to prevent it happening. Jacques Monod made this point very well in his Herbert Spencer lecture, after wryly remarking: 'Another curious aspect of the theory of evolution is that everybody thinks he understands it!'

To return to the primeval soup, it must have become populated by stable varieties of molecule; stable in that either the individual molecules lasted a long time, or they replicated rapidly, or they replicated accurately. Evolutionary trends toward these three kinds of stability took place in the following sense: if you had sampled the soup at two different times, the later sample would have contained a higher proportion of varieties with high longevity/fecundity/copying-fidelity. This is essentially what a biologist means by evolution when he is speaking of living creatures, and the mechanism is the same – natural selection.

Should we then call the original replicator molecules 'living'? Who cares? I might say to you 'Darwin was the greatest man who has ever lived', and you might say 'No, Newton was', but I hope we would not prolong the argument. The point is that no conclusion of substance would be affected whichever way our argument was resolved. The facts of the lives and achievements of Newton and Darwin remain totally unchanged whether we label them 'great' or not. Similarly, the story of the replicator molecules probably happened something like the way I am telling it, regardless of whether we choose to call them 'living'. Human suffering has been caused because too many of us cannot grasp that words are only tools for our use, and that the mere presence in the dictionary of a word like 'living' does not mean it necessarily has to refer to something definite in the real world. Whether we call the early replicators living or not, they were the ancestors of life; they were our founding fathers.

The next important link in the argument, one which Darwin himself laid stress on (although he was talking about animals and plants, not molecules) is *competition*. The primeval soup was not capable of supporting an infinite number of replicator molecules. For one thing, the earth's size is finite, but other limiting factors must also have been important. In our picture of the replicator acting as a template or mould, we supposed it to be bathed in a soup rich in the small building block molecules necessary to make copies. But when the replicators became numerous, building blocks must have been used up at such a rate that they became a scarce and precious resource. Different varieties or strains of replicator must have competed for them. We have considered the factors which would have increased the numbers of favoured kinds of replicator. We can now see that less-favoured varieties must actually have become *less* numerous because of competition, and ultimately many of their lines must have gone extinct. There was a struggle for existence among replicator varieties. They did not know they were struggling, or worry about it; the struggle was conducted without any hard feelings, indeed without feelings of any kind. But they were struggling, in the sense that any miscopying which resulted in a new higher level of stability, or a new way of reducing the stability of rivals, was automatically preserved and multiplied. The process of improvement was cumulative. Ways of increasing stability and of decreasing rivals' stability became more elaborate and more efficient. Some of them may even have 'discovered' how to break up molecules of rival varieties chemically, and to use the building blocks so released for making their own copies. These proto-carnivores simultaneously obtained food and removed competing rivals. Other replicators perhaps discovered how to protect themselves, either chemically, or by building a physical wall of protein around themselves. This may have been how the first living cells appeared. Replicators began not merely to exist, but to construct for themselves containers, vehicles for their continued existence. The replicators which survived were the ones which built *survival machines* for themselves to live in. The first survival machines probably consisted of nothing more than a protective coat. But making a living got steadily harder as new rivals arose with better and more effective survival machines. Survival machines got bigger and more elaborate, and the process was cumulative and progressive.

Was there to be any end to the gradual improvement in the techniques and artifices used by the replicators to ensure their own continuance in the world? There would be plenty of time for improvement. What weird engines of self-preservation would the millennia bring forth? Four thousand million years on, what was to be the fate of the ancient replicators? They did not die out, for they are past masters of the survival arts. But do not look for them floating loose in the sea; they gave up that cavalier freedom long ago. Now they swarm in huge colonies, safe inside gigantic lumbering robots, sealed off from the outside world, communicating with it by tortuous indirect routes, manipulating it by remove control. They are in you and in me; they created us, body and mind; and their preservation is the ultimate rationale for our existence. They have come a long way, those replicators. Now they go by the name of genes, and we are their survival machines.

Rene Dubos

Extracts from *The Mirage of Health*, 1959

While modern science can boast of so many startling achievements in the health fields, its role has not been so unique and its effectiveness not so complete as is commonly claimed. In reality, as alrady stated, the monstrous spectre of infection had become but an enfeebled shadow of its former self by the time serums, vaccines, and drugs became available to combat microbes. Indeed, many of the most terrifying microbial diseases – leprosy, plague, typhus, and the sweating sickness, for example – had all but disappeared from Europe long before the advent of the germ theory. Similarly, the general state of nutrition began to improve and the size of children in the labour classes to increase even before 1900 in most of Europe and North America. The change became noticeable long before calories, balanced diets, and vitamins had become the pride of nutrition experts, the obsession of mothers, and a source of large revenues to the manufacturers of coloured packages for advertised food products.

Clearly, modern medical science has helped to clean up the mess created by urban and industrial civilization. However, by the time laboratory medicine came effectively into the picture the job had been carried far toward completion by the humanitarians and social reformers of the nineteenth century. Their romántic doctrine that nature is holy and healthful was scientifically naïve but proved highly effective in dealing with the most important health problems of their age. When the tide is receding from the beach it is easy to have the illusion that one can empty the ocean by removing water with a pail. The tide of infectious and nutritional diseases was rapidly receding when the laboratory scientist moved into action at the end of the past century. . . .

Life is like a large body of water moved by deep currents and by superficial breezes. We have gained some understanding of the winds and can adjust our sails to them. But the really powerful forces which determine population trends are deep currents of which we know little, the fundamental physical and biological laws of the world, the habits and beliefs of mankind with their roots deep in the past. It is intellectual deceit to be dogmatic in these matters on the basis of scientific knowledge, because information is so incomplete. And it is always dangerous to bring about radical and sudden social changes, because the complexity of the interrelationships in the living world inevitably makes for unforeseen consequences, often with disastrous results. The use of knowledge must be tempered by humility and common sense, and for this reason medical utopias must be taken with a great deal of salt. In this light it is important to remember that the control of microbial diseases in the Western world occurred progressively and was sufficiently slow to permit orderly adjustments. In contrast, the present efforts to eliminate infection rapidly in the underdeveloped countries by radical public health measures is almost bound to bring about biological disturbances and to give rise to new population problems before there has been time for achieving compensatory changes in the rest of the environment.

Of course, these considerations cannot influence the behaviour of the physician toward each individual patient. But they introduce new types of responsibilities in the problems of medical statesmanship involving whole populations. They illustrate how difficult it is to define and delineate the role of medicine in the community. In the words of a wise physician, it is part of the doctor's function to make it possible for his patients to go on doing the pleasant things that are bad for them – smoking too much, eating too much, drinking too much – without killing themselves any sooner than is necessary. But it is also the doctor's role, as claimed by Rudolf Virchow, to recognize that social disease is the manifestation of a process affecting the community as a whole. . . .

Freeman Dyson

'The argument from design', from *Disturbing the Universe*, 1979

Professional scientists today live under a taboo against mixing science and religion. This was not always so. When Thomas Wright, the discoverer of galaxies, announced his discovery in 1750 in his book *An Original Theory or New Hypothesis of the Universe*, he was not afraid to use a theological argument to support an astronomical theory:

> Since as the Creation is, so is the Creator also magnified, we may conclude in consequence of an infinity, and an infinite all-active power, that as the visible creation is supposed to be full of siderial systems and planetary worlds, so on, in like similar manner, the endless immensity is an unlimited plenum of creations not unlike the known universe. . . . That this in all probability may be the real case, is in some degree made evident by the many cloudy spots, just perceivable by us, as far without our starry Regions, in which tho' visibly luminous spaces, no one star or particular constituent body can possibly be distinguished; those in all likelyhood may be external creation, bordering upon the known one, too remote for even our telescopes to reach.

Thirty-five years later, Wright's speculations were confirmed by William Herschel's precise observations. Wright also computed the number of habitable worlds in our galaxy:

> In all together then we may safely reckon 170,000,000, and yet be much within compass, exclusive of the comets which I judge to be by far the most numerous part of the creation.

His statement about the comets is also correct, although he does not tell us how he estimated their number. For him the existence of so many habitable worlds was not just a scientific hypothesis but a cause for moral reflection:

> In this great celestial creation, the catastrophy of a world, such as ours, or even the total dissolution of a system of worlds, may possibly be no more to the great Author of Nature, than the most common accident in life with us, and in all probability such final and general Doomsdays may be as frequent there, as even Birthdays or mortality with us upon the earth. This idea has something so chearful in it, that I own I can never look upon the stars without wondering why the whole world does not become astronomers; and that men endowed with sense and reason should neglect a science they are naturally so much interested in, and so capable of inlarging the understanding, as next to a demonstration must convince them of their immortality, and reconcile them to all those little difficulties incident to human nature, without the least anxiety.
>
> All this the vast apparent provision in the starry mansions seem to promise: What ought we then not to do, to preserve our natural birthright to it and to merit such inheritance, which alas we think created all to gratify alone a race of vain-glorious gigantic beings, while they are confined to this world, chained like so many atoms to a grain of sand.

There speaks the eighteenth century. Now listen to the twentieth, speaking through the voices of the biologist Jacques Monod: 'Any mingling of knowledge with values is unlawful, forbidden', and of the physicist Steven Weinberg: 'The more the universe seems comprehensible, the more it also seems pointless.'

If Monod and Weinberg are truly speaking for the twentieth century, then I prefer the eighteenth. But in fact Monod and Weinberg, both of them first-rate scientists and leaders of research in their specialties, are expressing a point of view which does not take into account the subtleties and ambiguities of twentieth-century physics. The roots of their philosophical attitudes lie in the nineteenth century, not in the twentieth. The taboo against mixing knowledge with values arose during the nineteenth century out of the great battle between the evolutionary biologists led by Thomas Huxley and the churchmen led by Bishop Wilberforce. Huxley won the battle, but a hundred years later Monod and Weinberg were still fighting the ghost of Bishop Wilberforce.

The nineteenth-century battle revolved around the validity of an old argument for the existence of God, the argument from design. The argument from design says simply that the existence of a watch implies the existence of a watchmaker. Thomas Wright accepted this argument as valid in the astronomical domain. Until the nineteenth century, churchmen and scientists agreed that it was also valid in the domain of biology. The penguin's flipper, the nest-building instinct of the swallow, the eye of the hawk, all declare, like the stars and the planets in Addison's eighteenth-century hymn, 'The hand that made us is divine.' Then came Darwin and Huxley, claiming that the penguin and the swallow and the hawk could be explained by the process of natural selection operating on random hereditary variations over long periods of time. If Darwin and Huxley were right, the argument from design was demolished. Bishop Wilberforce despised the biologists, regarding them as irresponsible destroyers of faith, and fought them with personal ridicule. In public debate he asked Huxley whether he was descended from a monkey on his grandfather's or on his grandmother's side. The biologists never forgave him and never forgot him. The battle left scars which are still not healed.

Looking back on the battle a century later, we can see that Darwin and Huxley were right. The discovery of the structure and function of DNA has made clear the nature of the hereditary variations upon which natural selection operates. The fact that DNA patterns remain stable for millions of years, but are still occasionally variable, is explained as a consequence of the laws of chemistry and physics. There is no reason why natural selection operating on these patterns, in a species of bird that has acquired a taste for eating fish, should not produce a penguin's flipper. Chance variations, selected by the perpetual struggle to survive, can do the work of the designer. So far as the biologists are concerned, the argument from design is dead. They won their battle. But unfortunately, in the bitterness of their victory over their clerical opponents, they have made the meaninglessness of the universe into a new dogma. Monod states this dogma with his customary sharpness:

> The cornerstone of the scientific method is the postulate that nature is objective. In other words, the *systematic* denial that true knowledge can be got at by interpreting phenomena in terms of final causes, that is to say, of purpose.

Here is a definition of the scientific method that would exclude Thomas Wright from science altogether. It would also exclude some of the most lively areas of modern physics and cosmology.

It is easy to understand how some modern molecular biologists have come to accept a narrow definition of scientific knowledge. Their tremendous successes were achieved by reducing the complex behavior of living creatures to the simpler behavior of the molecules out of which the creatures are built. Their whole field of science is based on the reduction of the complex to the simple, reduction of the apparently purposeful movements of an organism to purely mechanical movements of its constituent parts. To the molecular biologist, a cell is a chemical machine, and the protein and nucleic acid molecules that control its behavior are little bits of clockwork, existing in well-defined states and reacting to their environment by changing from one state to another. Every student of molecular biology learns his trade by playing with models built of plastic balls and pegs. These models are an indispensable tool for detailed study of the structure and function of nucleic acids and enzymes. They are, for practical purposes, a useful visualization of the molecules out of which we are built. But from the point of view of a physicist, the models belong to the nineteenth century. Every physicist knows that atoms are not really little hard balls. While the molecular biologists were using these mechanical models to make their spectacular discoveries, physics was moving in a quite different direction.

For the biologists, every step down in size was a step toward increasingly simple and mechanical behavior. A bacterium is more mechanical than a frog, and a DNA molecule is more mechanical than a bacterium. But twentieth-century physics has shown that further reductions in size have an opposite effect. If we divide a DNA molecule into its component atoms, the atoms behave less mechanically than the molecule. If we divide an atom into nucleus and electrons, the electrons are less mechanical than the atom. There is a famous experiment, originally suggested by Einstein, Podolsky and Rosen in 1935 as a thought experiment to illustrate the difficulties of quantum theory, which demonstrates that the notion of an electron existing in an objective state independent of the experimenter is untenable. The experiment has been done in various ways with various kinds of particles, and the results show clearly that the state of a particle has a meaning only when a precise procedure for observing the state is prescribed. Among physicists there are many different philosophical viewpoints, and many different ways of interpreting the role of the observer in the description of subatomic processes. But all physicists agree with the experimental facts which make it hopeless to look for a description independent of the mode of observation. When we are dealing with things as small as atoms and electrons, the observer or experimenter cannot be excluded from the description of nature. In this domain, Monod's dogma, 'The cornerstone of the scientific method is the postulate that nature is objective', turns out to be untrue.

If we deny Monod's postulate, this does not mean that we deny the achievements of molecular biology or support the doctrines of Bishop Wilberforce. We are not saying that chance and the mechanical rearrangement of molecules cannot turn ape into man. We are saying only that if as physicists we try to observe in the finest detail the behavior of a single molecule, the meaning of the words 'chance' and 'mechanical' will depend upon the way we make our observations. The laws of subatomic physics cannot even be formulated without some reference to the observer. 'Chance' cannot be defined except as a measure of the observer's ignorance of the future. The laws leave a place for mind in the description of every molecule.

It is remarkable that mind enters into our awareness of nature on two separate levels. At the highest level, the level of human consciousness, our minds are somehow directly aware of the complicated flow of electrical and chemical patterns in our brains. At the lowest level, the level of single atoms and electrons, the mind of an observer is again involved in the description of events. Between lies the level of molecular biology, where mechanical models are adequate and mind appears to be irrelevant. But I, as a physicist, cannot help suspecting that there is a logical connection between the two ways in which mind appears in my universe. I cannot help thinking that our awareness of our own brains has something to do with the process which we call 'observation' in atomic physics. That is to say, I think our consciousness is not just a passive epiphenomenon carried along by the chemical events in our brains, but is an active agent forcing the molecular complexes to make choices between one quantum state and another. In other words, mind is already inherent in every electron, and the processes of human consciousness differ only in degree but not in kind from the processes of choice between quantum states which we call 'chance' when they are made by electrons.

Jacques Monod has a word for people who think as I do and for whom he reserves his deepest scorn. He calls us 'animists', believers in spirits. 'Animism', he says, 'established a covenant between nature and man, a profound alliance outside of which seems to stretch only terrifying solitude, Must we break this tie because the postulate of objectivity requires it?' Monod answers yes: 'The ancient covenant is in pieces; man knows at last that he is alone in the universe's unfeeling immensity, out of which he emerged only by chance.' I answer no. I believe in the covenant. It is true that we emerged in the universe by chance, but the idea of chance is itself only a cover for our ignorance. I do not feel like an alien in this universe. The more I examine the universe and study the details of its architecture, the more evidence I find that the universe in some sense must have known that we were coming.

There are some striking examples in the laws of nuclear physics of numerical accidents that seem to conspire to make the universe habitable. The strength of the attractive nuclear forces is just sufficient to overcome the electrical repulsion between the positive charges in the nuclei of ordinary atoms such as oxygen or iron. But the nuclear forces are not quite strong enough to bind together two protons (hydrogen nuclei) into a bound system which would be called a diproton if it existed. If the nuclear forces had been slightly stronger than they are, the diproton would exist and almost all the hydrogen in the universe would have been combined into diprotons and heavier nuclei. Hydrogen would be a rare element, and stars like the sun, which live for a long time by the slow burning of hydrogen in their cores, could not exist. On the other hand, if the nuclear forces had been substantially weaker than they are, hydrogen could not burn at all and there would be no heavy elements. If, as seems likely, the evolution of life requires a star like the sun, supplying energy at a constant rate for billions of years, then the strength of nuclear forces had to lie within a rather narrow range to make life possible.

A similar but independent numerical accident appears in connection with the weak interaction by which hydrogen actually burns in the sun. The weak interaction is millions of times weaker than the nuclear force. It is just weak enough so that the hydrogen in the sun burns at a slow and steady rate. If the weak interaction were much stronger or much weaker, any forms of life dependent on sunlike stars would again be in difficulties.

The facts of astronomy include some other numerical accidents that work to our advantage. For example, the universe is built on such a scale that the average distance between stars in an average galaxy like ours is about twenty million million miles, an extravagantly large distance by human standards. If a scientist asserts that the stars at these immense distances have a decisive effect on the possibility of human existence, he will be suspected of being a believer in astrology. But it happens to be true that we could not have survived if the average distance between stars were only two million million miles instead of twenty. If the distances had been smaller by a factor of ten, there would have been a high probability that another star, at some time during the four billion years that the earth has existed, would have passed by the sun close enough to disrupt with its gravitational field the orbits of the planets. To destroy life on earth, it would not be necessary to pull the earth out of the solar system. It would be sufficient to pull the earth into a moderately eccentric elliptical orbit.

All the rich diversity of organic chemistry depends on a delicate balance between electrical and quantum-mechanical forces. The balance exists only because the laws of physics include an 'exclusion principle' which forbids two electrons to occupy the same state. If the laws were changed so that electrons no longer excluded each other, none of our essential chemistry would survive. There are many other lucky accidents in atomic physics. Without such accidents, water could not exist as a liquid, chains of carbon atoms could not form complex organic molecules, and hydrogen atoms could not form breakable bridges between molecules.

I conclude from the existence of these accidents of physics and astronomy that the universe is an unexpectedly hospitable place for living creatures to make their home in. Being a scientist, trained in the habits of thought and language of the twentieth century rather than the eighteenth, I do not claim that the architecture of the universe proves the existence of God. I claim only that the architecture of the universe is consistent with the hypothesis that mind plays an essential role in its functioning.

We had earlier found two levels on which mind manifests itself in the description of nature. On the level of subatomic physics, the observer is inextricably involved in the definition of the objects of his observations. On the level of direct human experience, we are aware of our own minds, and we find it convenient to believe that other human beings and animals have minds not altogether unlike our own. Now we have found a third level to add to these two. The peculiar harmony between the structure of the universe and the needs of life and intelligence is a third manifestation of the importance of mind in the scheme of things. This is as far as we can go as scientists. We have evidence that mind is important on three levels. We have no evidence for any deeper unifying hypothesis that would tie these three levels together. As individuals, some of us may be willing to go further. Some of us may be willing to entertain the hypothesis that there exists a universal mind or world soul which underlies the manifestations of mind that we observe. If we take this hypothesis seriously, we are, according to Monod's definition, animists. The existence of a world soul is a question that belongs to religion and not to science.

When my mother was past eighty-five, she could no longer walk as she once did. She was restricted to short outings close to her home. Her favorite walk in those years was to a nearby graveyard which commands a fine view of the ancient city of Winchester and the encircling hills. Here I often walked with her and listened to her talk cheerfully of her approaching death. Sometimes, contemplating the stupidities of mankind, she became rather fierce. 'When I look at this world now,' she said once, 'it looks to me like an anthill with too many ants scurrying around. I think perhaps the best thing would be to do away with it altogether.' I protested, and she laughed. No, she said, no matter how enraged she was with the ants, she would never be able to do away with the anthill. She found it far too interesting.

Sometimes we talked about the nature of the human soul and about the Cosmic Unity of all souls that I had believed in so firmly when I was fifteen years old. My mother did not like the phrase Cosmic Unity. It was too pretentious. She preferred to call it a world soul. She imagined that she was herself a piece of the world soul that had been given freedom to grow and develop independently so long as she was alive. After death, she expected to merge back into the world soul, losing her personal identity but preserving her memories and her intelligence. Whatever knowledge and wisdom she had acquired during her life would add to the world soul's store of knowledge and wisdom. 'But how do you know that the world soul will want you back?' I said. 'Perhaps, after all these years, the world soul will find you too tough and indigestible and won't want to merge with you.' 'Don't worry about that', my mother replied. 'It may take a little while, but I'll find my way back. The world soul can do with a bit more brains.'

Otto R. Frisch

'Energy from nuclei', from What Little I Remember, 1979

Did I say that all nuclei had weights that were multiples of that of a hydrogen nucleus? That is not quite true; most of them are about 1% lighter than that, and therein lies the secret of nuclear (often called 'atomic') energy. When protons come together to form heavier nuclei their joint mass becomes less by an amount m, and a lot of energy E is set free, following Einstein's formulae $E = mc^2$. The factor c^2 (speed of light multiplied by itself) is very large, so a minute amount of mass corresponds to a lot of energy; for instance the mass of a paper-clip is equivalent to the entire energy a small town uses during a day.

Energy is measured in a variety of units: kWh (kilowatt-hours) on your electricity meter, Btu (British thermal units) for the gas man, and so on. Those are man-size units, much too large for a single nucleus. For them the common unit is the MeV (a million electron-volt, but usually we say 'an emmeevee'). It is the energy of motion which an electron (or a proton) acquires when it is accelerated by a voltage of a million volts. An alpha particle has typically 5 to 10 MeV; to keep a watch going needs several million times as much energy every second.

Einstein's formula was put to the test in the 1930s by measuring the energy of the particles (e.g. protons) set free in 'atom splitting'. The collision of two nuclei caused the neucleons to be rearranged so as to form two new nuclei; when those were both of a kind found in nature it was possible to compare the masses of the nuclei before and after the collision and check the mass difference against the energy set free. That was done with mass spectrographs which were soon made so precise that Einstein's formula could be checked to within a fraction of an MeV; it was always found correct when the nuclei formed by the reaction were stable and hence available for mass spectroscopy. When unstable nuclei were formed one had to take into account the energy of the particles they subsequently sent out in transforming themselves into stable nuclei again. Soon there was a network of literally thousands of measurements, cross-checking each other, and the masses of several hundred isotopes were accurately known.

What do those masses tell us? Well, for one thing, they tell us why the sun keeps shining. If you could dive into the huge white-hot ball of not-quite-pure hydrogen which we call the sun you would find rapidly rising pressure and temperature until near the centre the temperature is around ten million degrees Centigrade. At such heat the hydrogen nuclei move so fast (about 500 km/sec) that they occasionally collide despite their mutual electric repulsion. There are traces of other elements, which complicate what happens; Hans Bethe, whom I later met in Los Alamos, was the first to work out a possible mechanism for this process in detail. To cut the story short, the main outcome is simply that helium nuclei are formed, one from four hydrogen nuclei (two of which are changed from protons into neutrons), and each hydrogen nucleus gives up 7 MeV in that process. In this 'nuclear fire' about a million times more energy is produced than in ordinary (chemical) fire, for instance when hydrogen burns by combining with oxygen. Even so the amount of hydrogen the sun has to burn to keep shining is stupendous: about ten billion tons every second! But the sun is big: in the four billion years since the Earth became solid the sun has used up only a fraction of its hydrogen.

If you go on to build up heavier nuclei you still liberate energy, but much less, and stars that run out of hydrogen become unstable. That raises fascinating questions regarding the nature of novae, supernovae, pulsars and so on. But here I'm getting on thin ice (or into hot water?), so let us return to solid ground.

Here we have some simple clues. Light nuclei contain as many neutrons as protons. The reason is a variant of Pauli's housing rule: two protons, spinning oppositely, can inhabit one quantum state, together with two neutrons behaving the same way. The first complete family of that kind is indeed the helium nucleus, rare on Earth but exceedingly common in the sun and the stars. But then why do heavier nuclei contain relatively more neutrons? Why is the ratio of neutrons to protons about 1.2:1 for copper, 1.4:1 for iodine and 1.6:1 for uranium? Because protons are bad club members: they are electrically charged and hence repel each other, and it makes a heavy nucleus more stable if some of them are turned into neutrons even though, as a result, they may have to move into higher quantum states. Nuclei with too few or too many protons adjust the ratio after a while by sending out an electron or a positron, as I mentioned earlier.

But in the heaviest nuclei, even when the ratio of neutrons to protons is at its optimum, the protons are still under pressure from their mutual repulsion. Then why don't they just get pushed out? In fact what is holding nuclei together? The protons repel each other, and the neutrons – being electrically neutral – cannot be held by electric forces. Gravity? Many million times too weak.

Today we know that any two nucleons attract each other very strongly, but only when they are very close together. We have no special name for that attraction; we call it simply 'the nuclear force'. It is more like a kind of stickiness, and we even think we know something about the nature of the glue. It acts only between nucleons in the same nucleus, except for a brief moment when two nuclei collide.

But the heavy nuclei have a trick to unload some of their quarrelsome protons. Two protons can combine with two neutrons and emigrate as a family; the 28 MeV which are gained (as in the process that keeps the sun shining!) serve to pay for the exit visa, as it were. In classical mechanics such a process would be impossible; like mountaineers trying to climb out of a crater on an insufficient supply of food, they would find that their energy gives out before they reach the rim and overcome the pull of the other nucleons.

Classical physics is adamant about that, but the laws of quantum mechanics are more flexible. They allow our subatomic mountaineer to 'tunnel' through the crater wall, as some physicists like to put it. Or you may imagine that two protons and two neutrons use Heisenberg's uncertainty principle to borrow some energy, to be repaid after they have left the nucleus and become a helium nucleus, a newborn alpha particle, rapidly driven away by the electric repulsion of the remaining nucleus, sliding down the outer crater wall as it were. But such a loan is granted only after uncounted billions of applications; in other words the chance of an alpha particle to escape in any given split second is minute and depends of course on the kind of nucleus. That chance was calculated from Schrödinger's wave equation, by Edward Condon (USA) with Ronald Gurney (UK), and also by the Russian, George Gamov, in 1926.

Until 1938 nobody dreamt that there was yet another way for a heavy nucleus to react to the mutual repulsion of its many protons, namely by dividing itself into two roughly equal halves. It was mere chance that I became involved in the discovery of that 'nuclear fission', which for the first time showed a way to make huge numbers of nuclei give up their hidden energy; the way to the atom bomb and to atomic power.

The occupation of Austria in March 1938 changed my aunt, the physicist Lise Meitner – technically – from an Austrian into a German. She had acquired fame by many years' work in Germany, but now had to fear dismissal as a descendant of a Jewish family. Moreover, there was a rumour that scientists might not be allowed to leave Germany; so she was persuaded – or perhaps stampeded – into leaving at very short notice, assisted by friends in Holland, and in the autumn she accepted an invitation to work in Stockholm, at the Nobel Institute led by Manne Siegbahn. I had always kept the habit of celebrating Christmas with her in Berlin; this time she was invited to spend Christmas with Swedish friends in the small town of Kungälv (near Gothenburg), and she asked me to join her there. That was the most momentous visit of my whole life.

Let me first explain that Lise Meitner had been working in Berlin with the chemist Otto Hahn for about thirty years, and during the last three years they had been bombarding uranium with neutrons and studying the radioactive substances that were formed. Fermi, who had first done that, thought he had made 'transuranic' elements – that is, elements beyond uranium (the heaviest element then known to the chemists), and Hahn the chemist was delighted to have a lot of new elements to study. But Lise Meitner saw how difficult it was to account for the large number of different substances formed, and things got even more complicated when some were found (in Paris) that were apparently lighter than uranium. Just before Lise Meitner left Germany, Hahn had confirmed that this was so, and that three of those substances behaved chemically like radium. It was hard to see how radium – four places below uranium – could be formed by the impact of a neutron, and Lise Meitner wrote to Hahn, imploring him not to publish that incomprehensible result until he was completely sure of it. Accordingly Hahn, together with his collaborator, the chemist Fritz Strassmann, decided to carry out thorough tests in order to make quite sure that those substances were indeed of the same chemical nature as radium.

When I came out of my hotel room after my first night in Kungälv I found Lise Meitner studying a letter from Hahn and obviously worried by it. I wanted to tell her of a new experiment I was planning, but she wouldn't listen; I had to read that letter. Its content was indeed so startling that I was at first inclined to be sceptical. Hahn and Strassmann had found that those three substances were not radium, chemically speaking; indeed they had found it impossible to separate them from the barium which, routinely, they had added in order to facilitate the chemical separations. They had come to the conclusion, reluctantly and with hesitation, that they were isotopes of barium.

Was it just a mistake? No, said Lise Meitner; Hahn was too good a chemist for that. But how could barium be formed from uranium? No larger fragments than protons or helium nuclei (alpha particles) had ever been chipped away from nuclei, and to chip off a large number not nearly enough energy was available. Nor was it possible that the uranium nucleus could have been cleaved right across. A nucleus was not like a brittle solid that can be cleaved or broken; George Gamov had suggested early on, and Bohr had given good arguments that a nucleus was much more like a liquid drop. Perhaps a drop could divide itself into two smaller drops in a more gradual manner, by first becoming elongated, then constricted, and finally being torn rather than broken in two? We knew that there were strong forces that would resist such a process, just as the surface tension of an ordinary liquid drop tends to resist its division into two smaller ones. But the nuclei differed from ordinary drops in one important way: they were electrically charged, and that was known to counteract the surface tension.

At that point we both sat down on a tree trunk (all that discussion had taken place while we walked through the wood in the snow, I with my skis on, Lise Meitner making good her claim that she could walk just as fast without), and started to calculate on scraps of paper. The charge of a uranium nucleus, we found, was indeed large enough to overcome the effect of the surface tension almost completely; so the uranium nucleus might indeed resemble a very wobbly, unstable drop, ready to divide itself at the slightest provocation, such as the impact of a single neutron.

But there was another problem. After separation, the two drops would be driven apart by their mutual electric repulsion and would acquire high speed and hence a very large energy, about 200 MeV in all; where could that energy come from? Fortunately Lise Meitner remembered the empirical formula for computing the masses of nuclei and worked out that the two nuclei formed by the division of a uranium nucleus together would be lighter than the original uranium nucleus by about one-fifth the mass of a proton. Now whenever mass disappears energy is created, according to Einstein's formula $E = mc^2$, and one-fifth of a proton mass was just equivalent to 200 MeV. So here was the source for that energy; it all fitted!

A couple of days later I travelled back to Copenhagen in considerable excitement. I was keen to submit our speculations – it wasn't really more at the time – to Bohr, who was just about to leave for the USA. He had only a few minutes for me; but I had hardly begun to tell him when he smote his forehead with his hand and exclaimed: 'Oh what idiots we all have been! Oh but this is wonderful! This is just as it must be! Have you and Lise Meitner written a paper about it?' Not yet, I said, but we would at once; and Bohr promised not to talk about it before the paper was out. Then he went off to catch his boat.

The paper was composed by several long-distance telephone calls, Lise Meitner having returned to Stockholm in the meantime. I asked an

American biologist who was working with Hevesy what they call the process by which single cells divide in two; 'fission', he said, so I used the term 'nuclear fission' in that paper. Placzek was sceptical; couldn't I do some experiments to show the existence of those fast-moving fragments of the uranium nucleus? Oddly enough that thought hadn't occurred to me, but now I quickly set to work, and the experiment (which was really very easy) was done in two days, and a short note about it was sent off to *Nature* together with the other note I had composed over the telephone with Lise Meitner. This time – with no Blackett to speed things up – about five weeks passed before *Nature* printed those notes.

In the meantime the paper by Hahn and Strassmann arrived in the USA, and several teams did within hours the same experiment which I had done on Placzek's challenge. A few days later Bohr heard about my own experiments, not from me (I wanted to get more results before wasting money on a transatlantic telegram!) but from his son Hans to whom I had casually talked about my work. Bohr responded with a barrage of telegrams, asking for details and proposing further experiments, and he worked hard to convince journalists that the decisive experiment had been done by Frisch in Copenhagen before the Americans. That was probably the source of the story – reprinted several times – that I was Bohr's son-in-law (although he never had a daughter, and I was then unmarried). I can see how it happened: a journalist asks: 'How do you know of this, Dr Bohr?' Bohr: 'My son wrote to me', Journalist mutters: 'His son, but name is Frisch; must be son-in-law'.

During this turmoil in the USA we were quietly continuing our work in Copenhagen. Lise Meitner felt that probably most of the radioactive substances which had been thought to lie beyond uranium – those 'transuranic' substances which Hahn thought they had discovered – were also fission products; a month or two later she came to Copenhagen and we proved that point by using a technique of 'radioactive recoil' which she had been the first to use, about thirty years previously. Yet transuranic elements were also formed; that was proved in California by Ed McMillan, with techniques much more sensitive than those available to Hahn and Meitner.

In all this excitement we had missed the most important point: the chain reaction. It was Christian Møller, a Danish colleague, who first suggested to me that the fission fragments (the two freshly formed nuclei) might contain enough surplus energy each to eject a neutron or two; each of these might cause another fission and generate more neutrons. By such a 'chain reaction' the neutrons would multiply in uranium like rabbits in a meadow! My immediate answer was that in that case no uranium ore deposits could exist: they would have blown up long ago by the explosive multiplication of neutrons in them. But I quickly saw that my argument was too naive; ores contained lots of other elements which might swallow up the neutrons; and the seams were perhaps thin, and then most of the neutrons would escape. So, from Møller's remark the exciting vision arose that by assembling enough pure uranium (with appropriate care!) one might start a controlled chain reaction and liberate nuclear energy on a scale that really mattered. Many others independently had the same thought, as I soon found out. Of course the spectre of a bomb – an uncontrolled chain reaction – was there as well; but for a while anyhow, it looked as though it need not frighten us. That complacency was based on an argument by Bohr, which was subtle but appeared quite sound.

In a paper on the theory of fission that he wrote in the USA with John Wheeler, Bohr concluded that most of the neutrons emitted by the fission fragments would be too slow to cause fission of the chief isotope, uranium–238. Yet slow neutrons did cause fission; this he attributed to the rare isotope uranium–235. If he was right the only chance of getting a chain reaction with natural uranium was to arrange for the neutrons to be slowed down, whereby their effect on uranium–235 is increased. But in that manner one could not get a violent explosion; slow neutrons take their time, and even if the conditions for rapid neutron multiplication were created this would at best (or at worst!) cause the assembly to heat up and disperse itself, with only a minute fraction of its nuclear energy liberated.

All this was quite correct, and the development of nuclear reactors followed on the whole the lines which Bohr foresaw. What he did not foresee was the fanatical ingenuity of the allied physicists and engineers, driven by the fear that Hitler might develop the decisive weapon before they did. I was in England when the war broke out, and in Los Alamos when I saw Bohr again. By that time it was clear that there were even two ways for getting an effective nuclear explosion: either through the separation of the highly fissile isotope uranium–235 or by using the new element plutonium, formed in a nuclear reactor. But I am again getting ahead of my story.

Francis Galton

An extract from his notes from his journeys in Africa

I profess to be a scientific man, and was exceedingly anxious to obtain accurate measurements of her shape; but there was a difficulty in doing this. I did not know a word of Hottentot, and could never therefore have explained to the lady what the object of my footrule could be; and I really dared not ask my worthy missionary host to interpret for me. The object of my admiration stood under a tree, and was turning herself about to all points of the compass, as ladies who wish to be admired usually do. Of a sudden my eye fell upon my sextant; the bright thought struck me, and I took a series of observations upon her figure in every direction, up and down, crossways, diagonally, and so forth, and I registered them carefully upon an outline drawing for fear of any mistake: this being done, I boldly pulled out my measuring tape, and measured the distance from where I was to the place where she stood, and having thus obtained both base and angles, I worked out the results by trigonometry and logarithms. . . .

William Gilbert

An extract from *De Magnete*, 1600

This natural philosophy (*physiologia*) is almost a new thing, unheard-of before; a very few writers have simply published some meagre accounts of certain magnetic forces. Therefore we do not at all quote the ancients and the Greeks as our supporters, for neither can paltry Greek argumentation demonstrate the truth more subtilly nor Greek terms more effectively, nor can both elucidate it better. Our doctrine of the loadstone is contradictory of most of the principles and axioms of the Greeks. Nor have we brought into this work any graces of rhetoric, any verbal ornateness, but have aimed simply at treating knotty questions about which little is known in such a style and in such terms as are needed to make what is said clearly intelligible. Therefore we sometimes employ words new and unheard-of, not (as alchemists are wont to do) in order to veil things with a pedantic terminology and to make them dark and obscure, but in order that hidden things which have no name and that have never come into notice, may be plainly and fully published. . . .

June Goodfield

An extract from An Imagined World

To understand the importance of Mrs. Wiggins and appreciate her efforts, it is best to go into the laboratory late at night. The janitors have long since been and gone, erasing the traces of the day's activity by the simple act of sweeping the floor and carrying out the garbage of science: scrawled notes, crunched-up diagrams, plastic coffee cups, wooden spatulas. The contents of the wastepaper basket are just as revealing as Henry Kissinger's ever were.

The floor may have been mopped but the benches are left untouched. The sinks will either be full of dirty glassware – to be dealt with by Mrs. Wiggins in the morning – or empty because the boys have already been in and carried the clutter away to the washer and the autoclave. There is not too much washing up anyway in these disposable days, and thus one traditional ritual of the scientist has vanished from the scene. There was a time, before the advent of technicians en masse, when collecting the dirty glassware and washing it at the end of the day was an ingrained habit of scientists, along with blowing their own glass pipettes and making up the biological solutions. But not now.

But by 11:00 p.m., all the Mrs. Wigginses have passed through and smoothed out the wrinkles of the earlier hours. The rooms are cleaner, the tidiness an orderly chaos, thoroughly familiar to those who work with it, because they have created both the order and the chaos and know just where to place their hands so as to find the stopwatch or the counters – given, of course, that no one else has removed them from their rightful places. The army of anonymous people who provide the infrastructure for science is thus as silent and remote as the scientists themselves. But they have left behind another army: serried rows of beakers, companies of flasks, battalions of tall cylinders, each topped with a silver-papered hat. The clean glassware in the cupboard and the sterile instruments in the drawers form the infantry of science, and all it takes to get them under the scientist's orders is an impersonal requisition form, an administrative signature or two, and, of course, the money to pay for it all. In Illinois or

Ohio, the glassware is molded and packed; from Maine or Minnesota, the sera and the chemical solutions are bottled and shipped; from California or Washington or London, the check that makes up the grant that underwrites the scientist who creates the edifice is signed and sent on its way, by a bureaucrat or a millionaire. Whether microscopes or Magic Markers, centrifuges or cells, patrons or administrators, the infrastructure of science is a vast array of people and things – in every respect an industry many orders of magnitude larger than the industry of science itself. The end result: the indifferent movement that without a moment's thought casually takes down a beaker from a shelf and marks the start of yet another experiment. . . .

J. E. Gordon

An extract from *The New Science of Strong Materials*, 1973

When wood begins to fail in compression little lines of buckled fibres can just be seen running at forty-five degrees to the grain but these are easily missed unless the surface is clean and you know what to look for. For some time after the initial failure nothing very sensational or catastrophic happens, the wood just yields gradually. In most cases wood is used in bending and the result of gradual crushing on the compression side of a beam is to transfer load to the tension side. In this way, the nominal stress in a wooden beam before actual collapse occurs may be up to twice the true compressive stress. It is this which makes a structure made out of timber such a safe one, generally one can very nearly get away with murder. Again, timber is noisy stuff and it will frighten the wits out of you before it is in any real danger of breaking. Sailplanes are often launched by means of half a mile or so of wire, reeled in by a winch. Having no engine, gliders are delightfully silent, except for a slight noise from the wind so that one can hear the structure very well. On a fast, gusty launch a wooden glider will treat you to a series of creaks and groans, and occasionally bangs, which are alarming until you realize that it is all pretence and that the structure is not in the least danger of breaking up. In fact it puts on this performance several times a day. I am pretty sure that these noises do not proceed from incipient compression failures. I have often wondered where they do come from but confess that I have absolutely no idea. They are however counted to wood for righteousness: as long as one can hear a timber structure one is very unlikely to break it.

For its weight, therefore, the strength of timber is as good or better than most of its competitors. Strength however is not enough: one must also have adequate stiffness. Substances like Nylon have plenty of strength but they are not sufficiently stiff to make engineering structures. The Young's modulus of spruce is about 1.5 to 2.0 $\times$ 10^6 lb. per square inch and the other timbers are, roughly, more or less stiff than this in proportion to their densities. Curiously, weight for weight, the Young's modulus of timbers is almost exactly the same as steel and aluminium and much better than synthetic resins. The good stiffness, combined with low density, means that wood is very efficient in beams and columns. Furniture, floors and bookshelves are usually best made in wood and so are things like flagstaffs and yacht's masts. The railways in America could be built very quickly and cheaply in the nineteenth century partly because of the efficiency of the timber trestle bridge.

As against these virtues, timber creeps. That is to say, if a stress is left on for a long time, wood will gradually run away from the load. This can be seen in the roof-tree of an old house or barn, which is generally concave. The creep of the wood is the reason why one must not leave a wooden bow or a violin tightly strung. The cause of the creep is most probably simply that, in the amorphous part of the cellulose, the rather badly stuck hydroxyls take advantage of changes in moisture and temperature to shuffle away from their responsibilities. It is unlikely that the crystalline part of cellulose creeps to any measurable extent.

An extract from *Structures*, 1978

In many respects the problems of persuading cloth to conform to a desired three-dimensional shape are not very different in sailmaking and in dressmaking. However, tailors and dressmakers seem to have been more intelligent about the matter than sailmakers. As far as was practicable they cut their cloth on the square, so that most of the circumferential or hoop stresses came directly along the line of the yarns. When a close fit was wanted it was achieved by what might be described as a system of Applied Tension: in other words, by lacing. At times the Victorian young lady seems to have had nearly as much rigging as a sailing ship.

With the virtual abandonment of systems of lacing in post-Edwardian times – possibly on account of a shortage of ladies' maids – women might well have had to face a shapeless future. However, in 1922 a dressmaker called Mlle Vionnet set up shop in Paris and proceeded to invent the 'bias cut'. Mlle Vionnet had probably never heard of her distinguished compatriot S. D. Poisson – still less of his ratio – but she realized intuitively that there are more ways of getting a fit than by pulling on strings or straining at hooks and eyes. The cloth of a dress is subject to vertical tensile stresses both from its own weight and from the movements of the wearer; and if the cloth is disposed at 45° to this vertical stress one can exploit the resulting large lateral contraction so as to get a clinging effect. The result was no doubt cheaper and more comfortable than the Edwardian solutions to the problem and, in selected instances, probably more devastating . . .

An analogous problem arises with the design of large rockets. Some rockets are driven by combinations of liquid fuels such as kerosene and liquid oxygen, but these systems involve elaborate plumbing which is liable to go wrong. Thus it may be better to use a 'solid' fuel such as that known as 'plastic propellant'. This stuff burns vigorously but relatively slowly, producing a great volume of hot gas which escapes through the rocket nozzle with a most impressive noise, driving the thing along as it does so. Both the propellant and the gas which it produces are contained within a strong cylindrical case or pressure vessel, whose walls must not be unduly exposed to flames or to high temperatures. For this reason the rather massive propellant charge is shaped in the form of a thick tube which fits tightly into the rocket casing. When the rocket is fired, combustion takes place at the inner surface of the plastic propellant, so that the tubular charge burns from the inside outwards. In this way the material of the case is protected from the flames up to the last possible moment by the presence of the remaining unburnt fuel.

Plastic propellant looks and feels rather like plasticine, and, like plasticine, it is apt to break in a brittle way, especially when it is cold. When a rocket is firing, the case naturally tends to expand under the gas pressure, rather as an artery expands under blood-pressure; if it does so, then the propellant has to expand with it. If the interior of the charge is still cold, it is likely to crack when the circumferential strain in the case reaches about 1.0 per cent. If this happens, then the flames will penetrate down the crack and destroy the case. This naturally results in a sensational explosion as another Polaris bites the dust.

Round about 1950, it occurred to some of us that it would be advantageous to make the rocket case, not from a metal tube, but in the form of a cylindrical vessel, wound from a double helix of strong glass fibres, bonded together with a resin adhesive. If the fibre angles are calculated correctly, it is possible so to arrange things that the change of diameter of the tube under pressure is small. It is true that, in such a situation, the tube will elongate more than it otherwise would, like Mlle Vionnet's waists, but, for various reasons, a longitudinal extension is less damaging to the propellant. As I seem to remember, this idea about rockets stemmed from the bias-cut nighties which were around at the time.

The strain requirements for rockets are generally just the opposite of what is needed in blood-vessels. . . . [O]ne wants an artery to maintain a constant length while exposed to fluctuations in blood-pressure (but changes in artery diameter are not important). Either condition can be met by making suitably designed tubes from helically disposed fibres. Problems of this kind keep cropping up in biology, and it was most interesting to find that Professor Steve Wainwright of Duke University, who is concerned with worms, has derived, quite independently, just the same mathematics as we had worked out twenty years or so before for use in rocketry. On inquiry, I find that in this case too the inspiration arose, via Professor Biggs, from the bias cut.

The invention of the bias cut brought fame to Mlle Vionnet in the world of haute couture. She lived to a great age and died, not long ago, at ninety-eight, quite unaware of her very significant contributions to space travel, to military technology and to the biomechanics of worms.

David Gould

An extract from *The Black and White Medicine Show*, 1985

Most doctors are not particularly interested in health. By inclination and training they are devoted to the study of disease. It is sick, not healthy people, who crowd their surgeries and out-patient departments and fill their hospital beds, and it is the fact that the population can be relied upon to provide a steady flow of sufferers from faults of the mind and of the flesh that guarantees them a job and an income in harsh times as in fair.

Not so long ago even the treatment (let alone the prevention) of disease took second place to diagnosis and prognosis, since so few effective remedies were available. The great physicians became renowned for their ability to pin an accurate label on their customers' complaints, and for their success in foretelling the outcome of a sickness, rather than for any real capacity to influence the course of events. The sections on treatment in medical textbooks, even as recently as 30 or 40 years ago, were commonly both brief and padded out with vague platitudes such as 'attention should be paid to the bowels', or 'adequate rest should be ensured by the administration of sedatives as required'. Doctors were valued (and valuable) for their priestly rather than their therapeutic skills.

Times have changed, so that now, too often, powerful and potentially harmful agents and techniques are employed in an effort to deal with patients' 'complaints' *before* a firm diagnosis has been reached, or before the likely benefits to the sufferer have been properly assessed.

Thus many a family doctor will prescribe penicillin for every customer presenting with a sore throat without considering the chances that the discomfort may well be due to a virus or some other factor that penicillin can't affect. Anxious patients are sent away with a prescription for a tranquillizer without an attempt to discover the cause of the anxiety, or whether it could be removed. Victims of advanced cancer are subjected to distressing procedures, and filled with drugs which cause them added misery, in a vain effort to halt the disease, when they ought to be nursed to an easy death.

The trouble is that today's doctor, far from being short of effectual therapeutic weapons, has a huge armamentarium of remedies, and since 'masterly inactivity' requires a confidence and strength of mind not universally possessed, over-treatment or inappropriate treatment, rather than lack of treatment is the hazard now faced by many of the customers who go to the medical trade for help.

Stephen Jay Gould

'Adam's navel', from *Natural History*, 1984

The ample fig leaf served our artistic forefathers well as a botanical shield against indecent exposure for Adam and Eve, our naked parents in the primeval bliss and innocence of Eden. Yet, in many ancient paintings, foliage hides more than Adam's genitalia; a wandering vine covers his navel as well. If modesty enjoined the genital shroud, a very different motive – mystery – placed a plant over his belly. In a theological debate more portentous than the old argument about angels on pinheads, many earnest people of faith had wondered whether Adam had a navel.

He was, after all, not born of a woman and required no remnant of his nonexistent umbilical cord. Yet, in creating a prototype, would not God make his first man like all the rest to follow? Would God, in other words, not create with the appearance of pre-existence? The issue was surely vexatious; in the absence of definite guidance, and not wishing to incur anyone's wrath, many painters literally hedged and covered Adam's belly.

A few centuries later, when the nascent science of geology was gathering evidence for the earth's enormous antiquity, some advocates of biblical literalism revived this old argument for our entire planet. The strata and their entombed fossils surely seem to represent a sequential record of countless years, but would not God create his earth with the appearance of pre-existence? Why should we not believe that he created strata and fossils to give modern life a harmonious order by granting it a sensible (if illusory) past? As God provide Adam with a navel to stress continuity with future men, so too did he endow a pristine world with the appearance of an ordered history. Thus, the earth might be but a few thousand years old, as Genesis literally affirmed, and still record an apparent tale of untold aeons.

This argument, so often cited as a premier example of reason at its most perfectly and preciously ridiculous, was most seriously and comprehensively set forth by the British naturalist Philip Henry Gosse in 1857. Gosse paid proper homage to the historical context of his argument in choosing a title for his volume. He named it *Omphalos* (Greek for navel), in Adam's honour, and added as a subtitle: *An Attempt to Untie the Geological Knot.*

Since *Omphalos* is such spectacular nonsense, readers may rightly ask why I choose to discuss it at all. I do so, first of all, because its author was such a serious and fascinating man, not a hopeless crank or malcontent. Any honest passion merits our attention, if only for the oldest of stated reasons – Terence's celebrated *Homo sum: humani nihil a me alienum puto* (I am human, and am therefore indifferent to nothing done by humans).

The David Attenborough of his day, Philip Henry Gosse (1810–88) was Britain's finest popular narrator of nature's fascination. He wrote a dozen books on plants and animals, lectured widely to popular audiences, and published several technical papers on marine invertebrates. He was also, in an age given to strong religious feeling as a mode for expressing human passions denied vent elsewhere, an extreme and committed fundamentalist of the Plymouth Brethren sect. Although his *History of the British Sea-Anemones* and other assorted ramblings in natural history are no longer read, Gosse retains some notoriety as the elder figure in that classic work of late Victorian self-analysis and personal exposé, his son Edmund's wonderful account of a young boy's struggle against a crushing religious extremism imposed by a caring and beloved parent – *Father and Son.*

My second reason for considering *Omphalos* invokes a common theme of mine: exceptions do prove rules (prove, that is, in the sense of probe or test, not affirm). If you want to understand what ordinary folks do, one thoughtful deviant will teach you more than ten thousand solid citizens.

When we grasp why *Omphalos* is so unacceptable (and not, by the way, for the reason usually cited), we will understand better how science and useful logic proceed. In any case, as an exercise in the anthropology of knowledge, *Omphalos* has no parallel – for its surpassing strangeness arose in the mind of a stolid Englishman, whose general character and cultural setting we can grasp as akin to our own, while the exotic systems of alien cultures are terra incognita both for their content and their context.

To understand *Omphalos*, we must begin with a paradox. The argument that strata and fossils were created all at once with the earth, and only present an illusion of elapsed time, might be easier to appreciate if its author had been an urban armchair theologian with no feeling or affection for nature's works. But how could a keen naturalist, who had spent days, nay months, on geological excursions, and who had studied fossils hour after hour, learning their distinctions and memorizing their names, possibly be content with the prospect that these objects of his devoted attention had never existed – were, indeed, a kind of grand joke perpetrated upon us by the Lord of All?

Philip Henry Gosse was the finest descriptive naturalist of his day. His son wrote: 'As a collector of facts and marshaller of observations, he had not a rival in that age.' The problem lies with the usual caricature of *Omphalos* as an argument that God, in fashioning the earth, had consciously and elaborately lied either to test our faith or simply to indulge in some inscrutable fit of arcane humour. Gosse, so fiercely committed both to his fossils and his God, advanced an opposing interpretation that commanded us to study geology with diligence and to respect all its facts even though they had no existence in real time. When we understand why a dedicated empiricist could embrace the argument of *Omphalos* ('creation with the appearance of preexistence'), only then can we understand its deeper fallacies.

Gosse began his argument with a central, but dubious, premise: All natural processes, he declared, move endlessly round in a circle: egg to chicken to egg, oak to acorn to oak.

> This, then, is the order of all organic nature. When once we are in any portion of the course, we find ourselves running in a circular groove, as endless as the course of a blind horse in a mill. . . . [In premechanized mills, horses wore blinders or, sad to say, were actually blinded, so that they would continue to walk a circular course and not attempt to move straight forward, as horses relying on visual cues tend to do.] This is not the law of some particular species, but of all: it pervades all classes of animals, all classes of plants, from the queenly palm down to the protococcus, from the monad up to man: the life of every organic being is whirling in a ceaseless circle, to

which one knows not how to assign any commencement. . . . The cow is as inevitable a sequence of the embryo, as the embryo is of the cow.

When God creates, and Gosse entertained not the slightest doubt that all species arose by divine fiat with no subsequent evolution, he must break (or 'irrupt', as Gosse wrote) somewhere into this ideal circle. Wherever God enters the circle (or 'places his wafer of creation', as Gosse stated in metaphor), his initial product must bear traces of previous stages in the circle, even if these stages had no existence in real time. If God chooses to create humans as adults, their hair and nails (not to mention their navels) testify to previous growth that never occurred. Even if he decides to create us as a simple fertilized ovum, this initial form implies a phantom mother's womb and two nonexistent parents to pass along the fruit of inheritance.

> Creation can be nothing else than a series of irruptions into circles. . . . Supposing the irruption to have been made at what part of the circle we please, and varying this condition indefinitely at will, – we cannot avoid the conclusion that each organism was from the first marked with the records of a previous being. But since creation and previous history are inconsistent with each other; as the very idea of the creation of an organism excludes the idea of pre-existence of that organism, or of any part of it; it follows, that such records are false, so far as they testify to time.

Gosse then invented a terminology to contrast the two parts of a circle before and after an act of creation. He labelled as 'prochronic', or occurring outside time, those appearances of pre-existence actually fashioned by god at the moment of creation but seeming to mark earlier stages in the circle of life. Subsequent events occurring after creation, and unfolding in conventional time, he called 'diachronic'. Adam's navel was prochronic; the 930 years of his earthly life were diachronic.

Gosse devoted more than 300 pages, some ninety percent of his text, to a simple list of examples for the following small part of his complete argument – if species arise by sudden creation at any point in their life cycle, their initial form must present illusory (prochronic) appearances of pre-existence. Let me choose just one among his numerous illustrations, both to characterize his style of argument and to present his gloriously purple prose. If God created vertebrates as adults, Gosse claimed, their teeth imply a prochronic past in patterns of wear and replacement.

Gosse leads us on an imaginary tour of life just an hour after its creation in the wilderness. He pauses at the seashore and scans the distant waves:

> I see yonder a . . . terrific tyrant of the sea . . . It is the grisly shark. How stealthily he glides along . . . Let us go and look into his mouth . . . Is not

STEPHEN JAY GOULD   71

this an awful array of knives and lancets? Is not this a case of surgical instruments enough to make you shudder? What would be the amputation of your leg to this row of triangular scalpels?

Yet the teeth grow in spirals, one behind the next, each waiting to take its turn as those in current use wear down and drop out:

> It follows, therefore, that the teeth which we now see erect and threatening, are the successors of former ones that have passed away, and that they were once dormant like those we see behind them. . . . Hence we are compelled by the phenomena to infer a long past existence to this animal, which yet has been called into being within an hour.

Should we try to argue that teeth in current use are the first members of their spiral, implying no predecessors after all, Gosse replies that their state of wear indicates a prochronic past. Should we propose that these initial teeth might be unmarred in a newly created shark, Gosse moves on to another example:

> Away to a broader river. Here wallows and riots the huge hippopotamus. What can we make of his dentition?

All modern adult hippos possess strongly worn and bevelled canines and incisors, a clear sign of active use throughout a long life. May we not, however, as for our shark, argue that a newly created hippo might have sharp and pristine front teeth? Gosse argues correctly that no hippo could work properly with teeth in such a state. A created adult hippo must contain worn teeth as witnesses of a prochronic past:

> The polished surfaces of the teeth, worn away by mutual action, afford striking evidence of the lapse of time. Some one may possibly object. . . . 'What right have you to assume that these teeth were worn away at the moment of its creation, admitting the animal to have been created adult. May they not have been entire?' I reply, Impossible: the Hippopotamus's teeth would have been perfectly useless to him, except in the ground-down condition: nay, the unworn canines would have effectually prevented his jaws from closing, necessitating the keeping of the mouth wide open until the attrition was performed; long before which, of course, he would have starved. . . . The degree of attrition is merely a question of time. . . . How distinct an evidence of past action, and yet, in the case of the created individual, how illusory!

This could go on for ever (in the book it nearly does), but just one more dental example. Gosse, continuing upward on the topographic trajectory of his imaginary journey, reaches an inland wood and meets *Babirussa*, the famous Asian pig with upper canines growing out and arching back, almost piercing the skull:

In the thickets of this nutmeg grove beside us there is a Babiroussa; let us examine him. Here he is, almost submerged in this tepid pool. Gentle swine with the circular tusk, please to open your pretty mouth!

The pig, created by God but an hour ago, obliges, thus displaying his worn molars and, particularly, the arching canines themselves, a product of long and continuous growth.

I find this part of Gosse's argument quite satisfactory as a solution, within the boundaries of his assumptions, to that classic dilemma of reasoning (comparable in importance to angels on pinheads and Adam's navel): 'Which came first, the chicken or the egg?' Gosse's answer: 'Either, at God's pleasure, with prochronic traces of the other.' But arguments are only as good as their premises, and Gosse's inspired nonsense fails because an alternative assumption, now accepted as undoubtedly correct, renders the question irrelevant – namely, evolution itself. Gosse's circles do not spin around eternally; each life cycle traces an ancestry back to inorganic chemicals in a primeval ocean. If organisms arose by acts of creation *ab nihilo*, then Gosse's argument about prochronic traces must be respected. But if organisms evolved to their current state, *Omphalos* collapses to massive irrelevance. Gosse understood this threat perfectly well and chose to meet it by abrupt dismissal. Evolution, he allowed, discredited his system, but only a fool could accept such patent nonsense and idolatry (Gosse wrote *Omphalos* two years before Darwin published the *Origin of Species*).

> If any choose to maintain, as many do, that species were gradually brought to their present maturity from humbler forms . . . he is welcome to his hypothesis, but I have nothing to do with it. These pages will not touch him.

But Gosse then faced a second and larger difficulty: the prochronic argument may work for organisms and their life cycles, but how can it be applied to the entire earth and its fossil record – for Gosse intended *Omphalos* as a treatise to reconcile the earth with biblical chronology: 'an attempt to untie the geological knot.' His statements about prochronic parts in organisms are only meant as collateral support for the primary geological argument. And Gosse's geological claim fails precisely because it rests upon such dubious analogy with what he recognized (since he gave it so much more space) as a much stronger argument about modern organisms.

Gosse tried valiantly to advance for the entire earth the same two premises that made his prochronic argument work for organisms. But an unwilling world rebelled against such forced reasoning and *Omphalos* collapsed under its own weight of illogic. Gosse first tried to argue that all geological processes, like organic life cycles, move in circles:

The problem, then, to be solved before we can certainly determine the question of analogy between the globe and the organism, is this – Is the life-history of the globe a cycle? If it is (and there are many reasons why this is probable), then I am sure prochronism must have been evident at its creation, since there is no point in a circle which does not imply previous points.

But Gosse could never document any inevitable geological cyclicity, and his argument drowned in a sea of rhetoric and biblical allusion from Ecclesiastes: 'All the rivers run into the sea; yet the sea is not full. Unto the place from whence the rivers come, thither they return again.'

Secondly, to make fossils prochronic, Gosse had to establish an analogy so riddled with holes that it would make the most ardent mental tester shudder – embryo is to adult as fossil is to modern organism. One might admit that chickens require previous eggs, but why should a modern reptile (especially for an antievolutionist like Gosse) be necessarily linked to a previous dinosaur as part of a cosmic cycle? A python surely does not imply the ineluctable entombment of an illusory *Triceratops* into prochronic strata.

With this epitome of Gosse's argument, we can resolve the paradox posed at the outset. Gosse could accept strata and fossils as illusory and still advocate their study because he did not regard the prochronic part of a cycle as any less 'true' or informative than its conventional diachronic segment. God decreed two kinds of existence – one formed all at once with the appearance of elapsed time, the other progressing sequentially. Both dovetail harmoniously to form uninterrupted circles that, in their order and majesty, give us insight into God's thoughts and plans.

The prochronic part is neither a joke nor a test of faith: it represents God's obedience to his own logic, given his decision to order creation in circles. As thoughts in God's mind, solidified in stone by creation *ab nihilo*, strata and fossils are just as true as if they recorded the products of conventional time. A geologist should study them with as much care and zeal, for we learn God's ways from both his prochronic and diachronic objects. The geological time-scale is no more meaningful as a yardstick than as a map of God's thoughts.

> The acceptance of the principles presented in this volume . . . would not, in the least degree, affect the study of scientific geology. The character and order of the strata; . . . the successive floras and faunas; and all the other phenomena, would be facts still. They would still be, as now, legitimate subjects of examination and inquiry . . . We might still speak of the inconceivably long duration of the processes in question, provided we understand ideal instead of actual time – that the duration was projected in the mind of God, and not really existent.

Thus, Gosse offered *Omphalos* to practising scientists as a helpful resolution of potential religious conflicts, not a challenge to their procedures or the relevance of their information.

His son Edmund wrote of the great hopes that Gosse held for *Omphalos*:

> Never was a book cast upon the waters with greater anticipations of success than was this curious, this obstinate, this fanatical volume. My father lived in a fever of suspense, waiting for the tremendous issue. This *Omphalos* of his, he thought, was to bring all the turmoil of scientific speculation to a close, fling geology into the arms of Scripture, and make the lion eat grass with the lamb.

Yet readers greeted *Omphalos* with disbelief, ridicule, or worse, stunned silence. Edmund Gosse continued:

> He offered it, with a glowing gesture, to atheists and Christians alike. This was to be the universal panacea; this the system of intellectual therapeutics which could not but heal all the maladies of the age. But, alas! atheists and Christians alike looked at it and laughed, and threw it away.

Although Gosse reconciled himself to a God who would create such a minutely detailed, illusory past, this notion was anathema to most of his countrymen. The British are a practical, empirical people, 'a nation of shopkeepers' in Adam Smith's famous phrase; they tend to respect the facts of nature at face value and rarely favour the complex systems of nonobvious interpretation so popular in much of continental thought. Prochronism was simply too much to swallow. The Reverend Charles Kingsley, an intellectual leader of unquestionable devotion to both God and science, spoke for a consensus in stating that he could not 'give up the painful and slow conclusion of five and twenty years' study of geology, and believe that God has written on the rocks one enormous and superfluous lie.'

And so it has gone for the argument of *Omphalos* ever since. Gosse did not invent it, and a few creationists ever since have revived it from time to time. But it has never been welcome or popular because it violates our intuitive notion of divine benevolence as free of devious behaviour – for while Gosse saw divine brilliance in the idea of prochronism, most people cannot shuck their seat-of-the-pants feeling that it smacks of plain old unfairness. Our modern American creationists reject it vehemently as imputing a dubious moral character to God and opt instead for the even more ridiculous notion that our miles of fossiliferous strata are all products of Noah's flood and can therefore be telescoped into the literal time scale of Genesis.

But what is so desperately wrong with *Omphalos*? Only this really (and perhaps paradoxically): that we have no way to find out whether it is wrong – or, for that matter, right. It is the classic example of an utterly untestable notion, for the world will look exactly the same in all its intricate detail whether fossils and strata are prochronic or products of an extended history. When we realize that *Omphalos* must be rejected for this methodological absurdity, not for any proved factual inaccuracy, then we will understand science as a way of knowing, and *Omphalos* will serve its purpose as an intellectual foil or prod.

Science is a procedure for testing and rejecting hypotheses, not a compendium of certain knowledge. Claims that can be proved incorrect lie within its domain (as false statements to be sure, but as proposals that meet the primary methodological criterion of testability). But theories that cannot be tested in principle are not part of science. Science is doing, not clever cogitation; we reject *Omphalos* as useless, not wrong.

Gosse's deep error lay in his failure to recognize this essential character of scientific reasoning. He hammered his own coffin nails by continually emphasizing that *Omphalos* made no practical difference – that the world would look exactly the same with a prochronic or diachronic past. (Gosse thought that this admission would make his argument acceptable to conventional geologists; he never realized that it could only lead them to reject his entire scheme as irrelevant.) 'I do not know,' he wrote, 'that a single conclusion, now accepted, would need to be given up, except that of actual chronology.'

Gosse emphasized that we cannot know where God placed his wafer of creation into the cosmic circle because prochronic objects, created *ab nihilo*, look exactly like diachronic development of actual time. To those who argued that coprolites (fossil excrement) prove the existence of active, feeding animals in a real geological past, Gosse replied that as God would create adults with faeces in their intestines, so too would he place petrified turds into his created strata (I am not making up this example for comic effect; you will find it on page 353 of *Omphalos*). Thus, with these words, Gosse sealed his fate and placed himself outside the pale of science:

> Now, again I repeat, there is no imaginable difference to sense between the prochronic and the diachronic development. Every argument by which the physiologist can prove to demonstration that yonder cow was once a foetus in the uterus of its dam, will apply with exactly the same power to show that the newly created cow was an embryo, some years before its creation. . . . There is, and can be, nothing in the phenomena to indicate a commencement there, any more than anywhere else, or indeed, anywhere at all. The commencement, as a fact, I must learn from testimony; I have no means whatever of inferring it from phenomena.

Gosse was emotionally crushed by the failure of *Omphalos*. During the long winter evenings of his discontent, in the January cold of 1858, he sat by the fire with his eight-year-old son, trying to ward off bitter thoughts by discussing the grisly details of past and current murders. Young Edmund heard of Mrs Manning, who buried her victim in quicklime and was hanged in black satin; of Burke and Hare, the Scottish ghouls; and of the 'carpetbag mystery', a sackful of neatly butchered human parts hung from a pier on Waterloo Bridge. This may not have been the most appropriate subject for an impressionable lad (Edmund was, by his own memory, 'nearly frozen into stone with horror'), yet I take some comfort in the thought that Philip Henry Gosse, smitten with the pain of rejection for his untestable theory, coudl take refuge in something so unambiguously factual, so utterly concrete.

Richard L. Gregory

'Blindness, recovery from', from *The Oxford Companion to the Mind*, 1987

What would an adult who had been blind all his life be able to see if the cause of his blindness was suddenly removed? This question was asked by several empiricist philosophers in the eighteenth century. John Locke considered the possibilities of such a case in 1690, following the question posed in a letter from his friend William Molyneux:

> Suppose a man born blind, and now adult, and taught by his touch to distinguish between a cube and a sphere of the same metal. Suppose then the cube and the sphere were placed on a table, and the blind man made to see: query, whether by his sight, before he touched them, could he distinguish and tell which was the globe and which the cube? . . . The acute and judicious proposer answers: not. For though he has obtained the experience of how the globe, how the cube, affects his touch, yet he has not yet attained the experience that what affects his touch so or so, must affect his sight, so or so. . . .

Locke comments in the *Essay concerning Human Understanding* (1690), Book II, Ch. 9, Sect. 8:

> I agree with this thinking gentleman, whom I am pleased to call my friend, in his answer to this problem; and am of the opinion that the blind man, at first, would not be able with certainty to say which was the globe, which the cube. . . .

René Descartes, in a passage in the *Dioptrics* (1637), considers how a blind man might build up a perceptual world by tapping around him with a stick. He first considers a sighted person using a stick in darkness. Descartes must surely have tried this for himself, and perhaps actually tested blind people. He says, of this experiment:

> without long practice this kind of sensation is rather confused and dim; but if you take men born blind, who have made use of such sensations all their life, you will find they feel things with perfect exactness that one might almost say that they see with their hands. . . .

Descartes goes on to suggest that normal vision resembles a blind man exploring and building up his sense world by successive probes with a stick.

George Berkeley, in *A New Theory of Vision* (1709), sect. lxxxv, stresses the importance of touch for seeing by considering that a microscope, by so changing the scale of things that touch no longer corresponds to vision, is of little use; and so, if our eyes were 'turned into the nature of microscopes, we should not be much benefited by the change . . . and (we would be) left only with the empty amusement of seeing, without any other benefit arising from it' (sect. lxxxvi). Berkeley goes on to say that we should expect a blind man who recovered sight not to know visually whether anything was

> high or low, erect or inverted . . . for the objects to which he had hitherto used to apply the terms up and down, high and low, were such as only affected or were in some way perceived by touch; but the proper objects of vision make a new set of ideas, perfectly distinct and different from the former, and which can in no sort make themselves perceived by touch. (sect. xcv)

These remained interesting speculations, until in 1728 an unusually expert and thoughtful surgeon, William Cheselden, reported such a clinical case. Though generally distinguished as a surgeon his achievements were especially ophthalmic operations for cataract (Cope, *William Cheselden*, 1953). In a celebrated case Cheselden gave sight to a boy aged thirteen or fourteen who was born with highly opaque cataracts. Cheselden reported that:

> When he first saw, he was so far from making any judgment of distances, that he thought all object whatever touched his eyes (as he expressed it) as what he felt did his skin, and thought no object so agreeable as those which were smooth and regular, though he could form no judgment of their shape, or guess what it was in any object that was pleasing to him: he knew not the shape of anything, nor any one thing from another, however different in shape or magnitude; but upon being told what things were, whose form he knew before from feeling, he would carefully observe, that he might know them again; and (as he said) at first learned to know, and again forgot a

thousand things in a day. One particular only, though it might appear trifling, I will relate: Having often forgot which was the cat, and which the dog, he was ashamed to ask; but catching the cat, which he knew by feeling, he was observed to look at her steadfastly, and then, setting her down, said, So, puss, I shall know you another time. He was very much surprised, that those things which he had liked best, did not appear most agreeable to his eyes, expecting those persons would appear most beautiful that he loved most, and such things to be most agreeable to his sight, that were so to his taste. We thought he soon knew what pictures represented, which were shewed to him, but we found afterwards we were mistaken; for about two months after he was couched, he discovered at once they represented solid bodies, when to that time he considered them only as party-coloured planes, or surfaces diversified with variety of paint; but even then he was no less surprised, expecting the pictures would feel like the things they represented, and was amazed when he found those parts, which by their light and shadow appeared now round and uneven, felt only flat like the rest, and asked which was the lying sense, feeling or seeing?

Being shewn his father's picture in a locket at his mother's watch, and told what it was, he acknowledged the likeness, but was vastly surprised; asking, how it could be, that a large face could be expressed in so little room, saying, it should have seemed as impossible for him, as to put a bushel of anything into a pint. At first he could bear but very little light, and the things he saw, he thought extremely large; but upon seeing things larger, those first seen he conceived less, never being able to imagine any lines beyond the bounds he saw; the room he was in, he said, he knew to be but part of the house, yet he could not conceive that the whole house could look bigger. Before he was couched, he expected little advantage from seeing, worth undergoing an operation for, except reading and writing; for he said, he thought he could have no more pleasure in walking abroad than he had in the garden, which he could do safely and readily. And even blindness, he observed, had this advantage, that he could go anywhere in the dark, much better than those who can see; and after he had seen, he did not soon lose this quality, nor desire a light to go about the house in the night. He said, every new object was a new delight; and the pleasure was so great, that he wanted words to express it; but his gratitude to his operator he could not conceal, never seeing him for some time without tears of joy in his eyes, and other marks of affection. . . . A year after first seeing, being carried upon Epsom Downs, and observing a large prospect, he was exceedingly delighted with it, and called it a new kind of seeing. And now being couched in his other eye, he says, that objects at first appeared large to this eye, but not so large as they did at first to the other; and looking upon the same object with both eyes, he thought it looked about twice as large as with the first couched eye only, but not double, that we can in any ways discover.

Evidently the sensory worlds of touch and vision were not so separate as at least Berkeley imagined they would be, and visual perception developed remarkably rapidly. This was discussed in terms of the Cheselden case by the materialist philosopher Julien Offray de La Mettrie, in his dangerously challenging book, *Natural History of the Soul* (1745), where he argues that only education received through the senses makes man man, and gives him what we call a soul, while no development of the mind outwards ever takes place. A few years later, in his better known *Man A Machine* (1748), he says:

> Nothing, as any one can see, is so simple as the mechanism of our education. Everything may be reduced to sounds or words that pass from the mouth of one through the ears of another into his brain. At the same moment, he perceives through his eyes the shape of the bodies of which these words are arbitrary signs.

The findings of the Cheselden case (which, though by no means the first, is the first at all adequately reported) are confirmed by some later cases, though in others the development of perception is painfully slow. R. Latta described a case of a successful operation for congenital cataract in 1904, which was broadly similar with almost immediately useful vision; but very often the eye takes a long time to settle down after a cataract operation. This may explain why so many of the historical cases described by M. von Senden (1932) showed such slow development. This is, however, a controversial matter. The Canadian psychologist Donald Hebb, in *The Organization of Behaviour* (1949), attributed the general slowness to see after operation as evidence that a very great deal of learning is needed. This, indeed, is now generally accepted, but it remains a question how far previously gained knowledge from exploratory touch, the other senses, and from the reports of sighted people helps new-found vision.

The case of a man who received corneal grafts when aged 52 (Gregory and Wallace, *Recovery from Early Blindness*, 1963) has the advantage over previous cases that a good retinal image is immediately available after corneal grafts as the eye is far less disturbed than by a cataract operation. In this case the patient, 'S. B.', could see virtually immediately things he already knew by touch; though for objects or features where touch had not been available a long, slow learning process was required, and in fact he never became perceptually normal. It was striking that he had immediate visual perception for things that absolutely must have been learned while he was blind, and could not have been known innately. Thus, from extensive experience of feeling the hands of his pocket watch, he was able, immediately, to tell the time visually. Perhaps even more striking, as a boy at the blind school S. B. had been taught to recognize by touch capital letters which were inscribed on special wooden plates. This was useful for reading street names, brass plates, and so on. Now it turned out that he could immediately read capital letters visually, though not lower-case letters, which he had not been taught by touch. These took months to learn to see. The general finding, of this case, was dramatic transfer of knowledge from touch to vision. So S. B. was not like a baby learning to see: he already knew a great deal from touch, and this was available for his newly functioning vision.

He had difficulties with shadows. When walking on steps on a sunny day he would quite often step on the shadow, sometimes falling. He was remarkably good at judging horizontal though not vertical distances. Thus while still in the hospital he could give the distance of chairs or tables almost normally; but when looking down from the window he thought the ground was at touching distance though the ward was several stories high. When he was shown various distortion illusions it was found that he was hardly affected by them: they were almost undistorted. And he did not experience the usual spontaneous depth-reversals of the Necker cube – which appeared to him flat. Indeed, like the Cheselden case, he had great difficulty seeing objects in pictures. Cartoons of faces meant nothing to him. He also found some things he loved ugly (including his wife and himself!) and he was frequently upset by the blemishes and imperfections of the visible world. He was fascinated by mirrors: they remained wonderful to the end of his life. As for most of the cases, though, S. B. became severely depressed, and felt more handicapped with his vision than when blind.

A dramatic and revealing episode occurred when he was first shown a lathe, at the London Science Museum, shortly after he left hospital. He had a long-standing interest in tools, and he particularly wanted to be able to use a lathe.

> We led him to the glass case, and asked him to tell us what was in it. He was quite unable to say anything about it, except he thought the nearest part was a handle. (He pointed to the handle of the transverse feed.) He complained that he could not see the cutting edge, or the metal being worked, or anything else about it, and appeared rather agitated. We then asked the Museum Attendant for the case to be opened, and S. B. was allowed to touch the lathe. The result was startling; he ran his hands deftly over the machine, touching first the transverse feed handle and confidently naming it 'a handle', and then on to the saddle, the bed and the head-stock of the lathe. He ran his hands eagerly over the lathe, with his eyes shut. Then he stood back a little and opened his eyes and said: 'Now that I've felt it I can see.' He then named many of the parts correctly and explained how they would work, though he could not understand the chain of four gears driving the lead screw.

Many of these observations have been confirmed by Valvo (*Sight Restoration after Long-Term Blindness*, 1971) in half a dozen cases of recovery from blindness by a remarkable operation: fitting acrylic lenses to eyes that never formed completely. Tissue rejection of the artificial lenses was prevented by placing them in a tooth, which was implanted as a buffer in the eye. One of these Italian patients, wearing a lens in a tooth in his eye, is a philosopher! It is now possible to implant artificial lenses without tissue rejection, so perhaps more cases of adult recovery from infant blindness will now appear.

J. B. S. Haldane

Extracts from *The Causes of Evolution*, 1932

The individuals belonging to a species differ to a greater or less extent. We can divide the causes of variation into those which operated before and during the life of the individual. We take that life as beginning with the fusion of the nuclei of the gametes which formed it, namely, the egg and the spermatozoon in most animals, the ovule and pollen grain in higher plants. (The organism produced by the fusion of the gametes is called a *zygote*.) In many plants and a few animals we can study the effects of nurture, i.e. causes operating during the life of the individual, almost apart from those of nature, i.e. causes operating earlier. When a plant or animal can be propagated vegetatively, the vegetative progeny of a single individual resemble one another to an extraordinary degree, and are called a clone. Thus all the Cox's Orange Pippins in the world are grafted from one seedling. The differences which exist between members of a clone are mainly due to environment and not to heredity. Thus Cox's Orange is a very different plant according as it is grafted on French Paradise, which gives a suburban garden bush, or Broad-leafed English Paradise, which gives an orchard tree, yet in a given environment it behaves in a predictable way. Even within a clone new types may appear (so-called bud-sports). These generally produce their like when vegetatively propagated. But with these exceptions, differences within a clone are not inherited. They are the best example of what is called fluctuating variability, due to differences of environment, not transmissible by inheritance, and therefore irrelevant for the problem of evolution.

You cannot propagate guinea-pigs by cuttings, but by many generations of inbreeding you can produce a line of guinea-pigs extraordinarily alike. After twenty or more generations of brother and sister mating there is no more resemblance between parent and offspring than between cousins. You do not abolish variation, and if you choose a piebald race it is easy to study it. The pattern is affected by environment, especially by the age of the mother, but these variations are not inherited (Wright, 1926). . . .

But the main objection to *élan vital* is that it is so very erratically distributed. That sturdy little creature, the limpet, has watched the legions of evolution thunder by for some three hundred million years without changing its shell form to any serious extent. And the usual course taken by an evolving line has been one of degeneration. It seems to me altogether probable that man will take this course unless he takes conscious control of his evolution within the next few thousand years. It may very well be that mind, at our level, is not adequate for such a task, probably on account of its emotional rather than intellectual deficiencies. If that is the case we are perhaps the rather sorry climax of evolution, and less can be said in favour of existence than many of us suppose.

If I were compelled to give my own appreciation of the evolutionary process as seen in a great group such as the Ammonites, where it is completed, I should say this: In the first place, it is very beautiful. In that beauty there is an element of tragedy. On the human time-scale the life of a plant or animal species appears as the endless repetition of an almost identical theme. On the time-scale of geology we recapture that element of uniqueness, of *Einemaligkeit*, which makes the transitoriness of human life into a tragedy. In an evolutionary line rising from simplicity to complexity, then often falling back to an apparently primitive condition before its end, we perceive an artistic unity similar to that of a fugue, or the life work of a painter of great and versatile genius like Picasso, who began with severe line drawing, passed through cubism, and is now, in the intervals between still more bizarre experiments, painting somewhat in the manner of Ingres. Possibly such artistic work gives us as good insight into the nature of the reality around us as any other human activity. To me at least the beauty of evolution is far more striking than its purpose. . . .

'Do continents move?' from *Science and Everyday Life*, 1941

So far in this series of articles I have dealt with matters on which the majority of scientific workers are agreed. To-day I describe a theory which is held by a minority of geologists, and on which I am not myself qualified to judge. Many of the scientific controversies of to-day are hard to explain in a short article. This one is easy, though, of course, the evidence on each side cannot be given.

Twenty years ago almost all geologists thought that, although the continents had moved up and down, and some mountains had been formed by folding, they had stayed in the same place, or nearly so, relative to one another. If London is 3,200 miles from New York to-day, this was assumed to have been so when the rocks beneath the cities were formed.

The first men to doubt this in a systematic way were two American geologists, F. B. Taylor and H. B. Baker in 1910 and 1911; but Wegener of Graz, who was killed while exploring Greenland in 1930, carried these ideas a good deal further, and now many geologists think that the continents have moved relative to one another and are still moving.

A glance at a map of the world shows that if America were moved eastward in a block, the Atlantic Ocean would be filled up, and the British Isles would be brought close to Newfoundland and New England, while West Africa fitted against the West Indies, Brazil into the Gulf of Guinea, and so on. Baker believes that they fitted together in this way in the past.

The rocks, especially the older rocks, at corresponding points on opposite sides of the Atlantic correspond very well, or so it is claimed. Thus the coal-fields of South Wales appear again in Pennsylvania, the diamond-bearing belts on the coast of Brazil correspond to those of South-West Africa.

If the same jigsaw theory is applied in the southern hemisphere, Australia, Antarctica and India can all be fitted on to Africa. If so some very strange facts are explained. At the time when the coal was being formed in England, tillite – that is to say, boulder clay now converted into soft rock – was being laid down in South America, South Africa, India and Australia.

Clay containing boulders is only produced by ice, and the theory holds that all these continents were then grouped round the South Pole, while coal measures were being formed in the tropics. If these continents were once united, a number of very queer facts about geographical distribution of animals are explained. For example, marsupials, like the kangaroo and opossum, are mainly found in Australia and New Guinea, but also in South America; *Peripatus*, a strange animal like a caterpillar which never becomes an insect, is found in New Zealand, Australia, South Africa and South America.

In their drift apart, these southern continents have made the Indian and South Atlantic Oceans, while the Pacific Ocean represents the remains of a much larger original ocean, against which the Americas are pressing, piling up mountain ranges as they move westwards. Wegener thought that all the continents once formed a single mass. Du Toit and others think that the southern continents have approached Europe and Asia. The pressure of Africa against Europe made the Alps, Balkans, and Atlas Mountains, whilst the northern part of India actually penetrated under what is now central Asia, and lifted up the huge plateau of Tibet.

Africa is still splitting to-day, along the rift valley in Kenya, and in other places, and it can be predicted that in the course of millions of years much of East Africa will split off from the main continent, as Arabia and Madagascar have already done. Determinations of the longitude of places in Greenland seem to show that it is moving away from Europe at 20 or 30 yards per year. If this is finally proved to be so, there is no doubt that the theory of drifting continents will have to be accepted.

But how can continents move like this? We now know that they are made of rocks which are a good deal lighter than those composing the ocean floors, and can be regarded as floating on them. The heavy rocks are thought to be softer, and to crumple or stretch as the continents move. But it is hard to imagine what can be the huge forces which move the continents.

Perhaps the most hopeful theory is that of Holmes, which attributes it to currents of molten rock under the continents, rising in the hot centre of the continents, moving outwards, and sinking where the rocks have been cooled down by the ocean. Many geologists dismiss these ideas as absurd, but at present a number are supporting the new theories, and what is to-day a heresy may be orthodox to-morrow.

Among the strongest adherents are some economic geologists, such as the American Association of Petroleum Geologists. When we remember that geology owes so much to Smith, an engineer who classified the English strata when supervising the excavation of canals over a century ago, and to those who carried out geological surveys primarily for economic purposes, we may be inclined to back the practical men against the theorists who say that their views are impossible.

'Cancer's a funny thing', from the New Statesman, 1964

I wish I had the voice of Homer
To sing of rectal carcinoma,
Which kills a lot more chaps, in fact,
Than were bumped off when Troy was sacked.
I noticed I was passing blood
(Only a few drops, not a flood).
So pausing on my homeward way
From Tallahassee to Bombay
I asked a doctor, now my friend,

To peer into my hinder end,
To prove or to disprove the rumour
That I had a malignant tumour.
They pumped in BaSO$_4$
Till I could really stand no more,
And, when sufficient had been pressed in,
They photographed my large intestine,
In order to decide the issue
They next scraped out some bits of tissue.
(Before they did so, some good pal
Had knocked me out with pentothal,
Whose action is extremely quick,
And does not leave me feeling sick.)
The microscope returned the answer
That I had certainly got cancer.
So I was wheeled into the theatre
Where holes were made to make me better.
One set is in my perineum
Where I can feel, but can't yet see 'em.
Another made me like a kipper
Or female prey of Jack the Ripper.
Through this incision, I don't doubt,
The neoplasm was taken out,
Along with colon, and lymph nodes
Where cancer cells might find abodes.
A third much smaller hole is meant
To function as a ventral vent:
So now I am like two-faced Janus
The only[1] god who sees his anus.

[1] In India there are several more
  With extra faces, up to four,
    But both in Brahma and in Shiva
    I own myself an unbeliever.

I'll swear, without the risk of perjury,
It was a snappy bit of surgery.
My rectum is a serious loss to me,
But I've a very neat colostomy,
And hope, as soon as I am able,
To make it keep a fixed time-table.
So do not wait for aches and pains

To have a surgeon mend your drains;
If he says 'cancer' you're a dunce
Unless you have it out at once,
For if you wait it's sure to swell,
And may have progeny as well.
My final word, before I'm done,
Is 'Cancer can be rather fun'.
Thanks to the nurses and Nye Bevan
The NHS is quite like heaven
Provided one confronts the tumour
With a sufficient sense of humour.
I know that cancer often kills,
But so do cars and sleeping pills;
And it can hurt one till one sweats,
So can bad teeth and unpaid debts.
A spot of laughter, I am sure,
Often accelerates one's cure;
So let us patients do our bit
To help the surgeons make us fit.

The poem, which was reprinted in a number of countries, brought great praise, caused great offence, and in some ways crystallises both Haldane's attitude to the world and the world's reaction. 'I certainly intended to help others as well as myself', he later wrote. 'I am a good enough Marxist to think that every poem should have a social function, though not a good enough one to think that it must. The main functions of my rhyme, to induce cancer patients to be operated on early and to be cheerful about it, would have been better served had it included the following lines after the fourth:

Yet, thanks to modern surgeon's skills,
It can be killed before it kills
Upon a scientific basis
In nineteen out of twenty cases.

Roy Herbert

'On first encountering sodium benzoate', from the *New Scientist*, 1978

Sodium benzoate! What a name to rouse the blood. It conjures up the burning sands of Africa, the magnificence of ancient Rome, the sexual secrets of the Orient, the mystery of the Middle East.

Well, it does for me. It conjures up razor blades as well, to be honest, though I tend not to emphasise that in case bathos sets in. I first ran up against sodium benzoate in the days of the Chemical Research Laboratory at Teddington. I'd gone down there to find out about corrosion and what they were doing about it, a prospect that filled me chock-full of tedium, garnished with misgiving. How, I thought, would I be able to conceal the boredom convincingly, how to think of intelligent questions? The problem was, you see, that all these razor blades were rusting in their wraps. Those old enough to remember will know how the slightly oily paper of a blade stuck to the fingers, revealing the blotchy rust underneath, as if the blade had already been bloodily used.

The answer was sodium benzoate. Impregnate the paper with it and the blade remained shining, Excalibur-like. Moreover, sodium benzoate could prevent the attack of antifreeze agents on the cooling systems of motor cars. Further, sodium benzoate vapour could stop exported goods arriving in Singapore or Montevideo with corrosion all over them like fur coats. The sodium benzoate age was round the corner. I got the impression that speakers in Hyde Park were calling daily for immense increase in the sodium benzoate production of this country. UK tops sodium benzoate league. What is sodium benzoate, asks judge. In no time I was staring at the weather in a Perspex box. Round and round went bits of metal, enduring sprays of water, blasts of heat and ultraviolet radiation. Gleaming were those protected by sodium benzoate, right enough. But the weather box was limited in appeal and so I asked what else was new in the struggle against corrosion. And the world opened up as I was led into two prefabricated huts, presided over by Kenneth Butlin, the head of the microbiology section. The door opened on the sands of Africa. For, at the time, there was a sulphur shortage. Butlin and his colleagues, enthusiastically studying the life and works of sulphate reducing bacteria, had realised that bacteria could produce sulphur and, in fact, were doing that in natural ponds in North Africa. Even the background was romantic. Fascinated, I read that sulphur normally came from domes in the US. *Domes?* In Xanadu did Kubla

Khan a stately sulphur dome decree. The degree of industrial progress of a country could be measured in terms of its sulphuric acid consumption. That's a little known fact that I have used since at many a dinner party. Unfortunately, several hosts have taken it as an oblique criticism of their wine and will listen to no explanation.

Anyway, off Butlin had gone to North Africa to study the sulphur manufacturing capabilities of the bacteria, had been photographed sitting on a Roman latrine and, rumour had it, taken sheeps' eyes in his stride, to coin a phrase, at a feast in a sheikh's tent. He had also found, to his delight, that he had run into some opposition to his floundering about in the bright red ponds. There was a steady trade in the water from them. Bottled and sold in Cairo, it was looked on as a certain aphrodisiac. The opposition had not drawn a curved knife or two, but stuck to disapproval, a testimony to Butlin's colossal charm and character. He came back with bottles of water and in a matter of months he and his team had a process going on a laboratory scale that produced sulphur. In what to me is a mysterious way, that was all there was to it. Success appeared to achieve nothing. Sulphur supplies went back to normal. But the whole affair has left me with a sneaky feeling that there is more than meets the eye in corrosion. A wise man will poke about a bit when he runs across it. There's many an exciting tale told round the camp fire about corrosion. At the end, the listeners stare into the embers and throw their cups of sodium benzoate into the night before turning in.

I dare say you don't believe me. How about this? In May 1896, there was a reception at the Hotel Cecil that 'will long be remembered in the circles of fashionable London for several unique occurrences, whose record will doubtless pass into the imperishable annals of time', according to *The Lady*. Hundreds of the nobs were there. 'Peers and potentates, actresses and actors, musicians, authors, artists, scientists and soldiers and most of them whose celebrity is world-wide.' One of the staggering attractions was a lift. The reporter says, 'Then, with some other kindred spirits, I started on a voyage of discovery in a lift, whose powers of "lifting" are absolutely breath-staying in their celerity.' At first I thought she meant 'breath-taking'. Then I decided that Freud was among the throng. For at the end, appears an advertisement. 'GOOD NEWS', it says. 'LADIES will be DELIGHTED to learn they need no longer suffer the annoyance of RUSTY Corset Busks. Ask for TANAFITE Busks and Side Steels. TANAFITE is a speciality in steel, guaranteed rust-resisting and highly resilient.' Sodium benzoate would have made the advertiser a fortune. Or perhaps he wrapped the corset busks in oily paper, thus making the corsets rust-proof and a fire risk.

A. V. Hill

An extract from 'The ethical dilemma of science', from *Nature*, 1952

The dilemma is this. All the impulses of decent humanity, all the dictates of religion and all the traditions of medicine insist that suffering should be relieved, curable diseases cured, preventable disease prevented. The obligation is regarded as unconditional: it is not permitted to argue that the suffering is due to folly, that the children are not wanted, that the patient's family would be happier if he died. All that may be so; but to accept it as a guide to action would lead to a degradation of standards of humanity by which civilization would be permanently and indefinitely poorer. . . .

Some might [take] the purely biological view that if men will breed like rabbits they must be allowed to die like rabbits. . . . Most people would still say no. But suppose it were certain now that the pressure of increasing population, uncontrolled by disease, would lead not only to widespread exhaustion of the soil and of other capital resources but also to continuing and increasing international tension and disorder, making it hard for civilization itself to survive: Would the majority of humane and reasonable people then change their minds? If ethical principles deny our right to do evil in order that good may come, are we justified in doing good when the foreseeable consequence is evil?

Fred Hoyle

An extract from *Frontiers of Astronomy*, 1955

Man's claim to have progressed far beyond his fellow animals must be supported, not by his search for food, warmth, and shelter (however ingeniously conducted) but by his penetration into the very fabric of the Universe. It is in the world of ideas and in the relation of his brain to the Universe itself, that the superiority of Man lies. The rise of Man may justly be described as an adventure in ideas.

[I am here] concerned with one of the chapters of this adventure. It is in some respects the most spectacular chapter – the one in which the large scale features of the Universe are beginning to be unfolded to us in all their majesty. But we cannot seek for grandeur at the outset, rather must we begin very modestly with the Earth itself.

Let's start with the Earth . . .

Half the Earth is lit by the Sun and the other half lies in shadow. Because of the rotation of the Earth we are constantly turning from shadow to Sun and from Sun to shadow: we experience the procession of night and day. The Earth is turning with respect to the Sun rather like a joint on a spit, although of course there is no material spit stuck through the Earth – we are turning freely in space, just as we are moving freely in space on our annual journey around the Sun.

In the past the Earth rotated considerably more rapidly than it does now: at the time of its origin the cycle of day and night may have been as short as 10 hours. The spin of the Earth must accordingly have been slowed down during the 4,000 million years or so that have elapsed since the early period of its life. The agency responsible for the braking action is known. It is just the twice-daily tides that are raised by the Moon and the Sun. The oceanic tides cause a frictional resistance when they impinge on the continental margins. This friction produces heat at the expense of the energy of rotation of the Earth, thereby slightly slowing the Earth's spin. In return for its effect on the Earth, the Moon experiences a force that pushes it gradually farther and farther away from us.

Formerly it was thought that the spin of the Earth has been slowing down continuously ever since the time of its formation, so that on this old view it just happens that we are living at the time when the spin has been braked down to 24 hours: it was thought that in the past the cycle of night and day took less than 24 hours, and that in the future it would take more. But a very recent theory, due to E. R. R. Holmberg, disagrees with this last step, disagrees that the cycle of day and night will ever take longer than 24 hours in the future.

Now since the braking effect of the oceanic tides is undoubtedly still operative, this view of Holmberg evidently demands that there shall be some compensating process tending to speed up the spin of the Earth. The substance of the new argument is that such a compensating speeding-up process does in fact exist. To understand how it operates let us first consider an analogy. Take a weight suspended from a spring, give the weight a pull downwards, and then let go. The system will start oscillating up and down. Now give the weight a small push downward during each oscillation. It will be found that, provided the weight is always pushed at the same stage of each oscillation, a quite violent motion will be built up. This is known as forcing an oscillation in resonance – 'forcing' because of the pushes and 'in resonance' because the pushes are adjusted to come at the same stage of each oscillation.

Now the atmosphere of the Earth oscillates up and down like the spring and the weight, the pressure in the air taking the part of the spring and the weight of the atmosphere acting, of course, as the weight. Not only this, but the atmosphere is pushed by the same forces as those that raise the oceanic tides. But the force due to the Moon, which is the more important in the raising of the oceanic tides, does not act in resonance with the oscillations of the atmosphere and consequently does not build up appreciable motions of the atmospheric gases. The somewhat weaker pushes due to the Sun do act in resonance with the atmosphere, however. The result is that very considerable up and down motions of the air are set up. These motions are accompanied by oscillations of pressure that can be detected on a sensitive barometer. The variations occur twice daily, just as the oceanic tides do. The pressure is found to be at a maximum about two hours before midday and about two hours before midnight. By a careful calculation it can be shown that this precedence of the atmospheric tides before midday and midnight causes the gravitational field of the Sun to put a twist on the Earth tending to speed it up. The strength of the twist can also be estimated. The very important result emerges that the twist is comparable with the slowing down effect of the oceanic tides, just as Holmberg's theory requires it to be.

It is important to realise that the speeding-up process need not exactly compensate all the time for the slowing-down effect of the oceanic tides. It is sufficient if the two processes compensate each other *on the average*, averages being calculated over say a time of 100,000 years. Indeed exact equality at all times is not to be expected for the reason that the slowing effect is likely to vary quite appreciably and quickly from one time to another. During the last ice-age for instance the slowing effect may well have been much less than it is today.

The theory is also favoured by another point, one that seems to me to be well nigh decisive; namely that the time of oscillation of the atmosphere and the time between successive pushes of the Sun on the atmosphere depend on entirely different considerations. The time for the oscillation depends on the temperature, density, and chemical nature of the air, whereas the time between successive pushes of the Sun depends on the rate of spin of the Earth. How comes it then that the two are so closely coincident?

In answering this question, Holmberg follows the older ideas to begin with. He supposes that at one time the Earth was rotating considerably faster than at present. There was then no resonance between the pushes of the Sun and the oscillations of the atmosphere. Consequently no strong oscillations were set up, so that the speeding-up process was inappreciable. The slowing-down effect of the oceanic tides therefore operated essentially unchecked, just as used to be supposed. But here now is the crucial point. As the Earth slowed to a day of 24 hours the pushes of the Sun gradually came into resonance with the oscillations of the atmosphere. So larger and larger motions of the air were built up, and the speeding-up process increased correspondingly. This went on until the speeding-up process came into average balance with the slowing effect of the oceanic tides. A state of balance has been operative ever since.

These ideas of Holmberg lead to other interesting consequences. It appears that the Earth must be spiralling very slowly inwards towards the Sun, and the Moon must be spiralling slowly outwards from the Earth. The change in the distance of the Earth from the Sun remains very small, but the change in the distance of the Moon from the Earth does not remain small. Given sufficient time the Moon will spiral so far away from the Earth that it will fall prey to the gravitational influence of the Sun. The Sun will pull it away from the Earth entirely so that it will no longer go circling around us, but will move independently around the Sun as a planet in its own right. This will happen when the slow spiralling that is going on all the time takes the Moon out from its present distance of nearly a quarter of million miles to a distance of about a million miles. Long before this stage is reached we shall unfortunately lose one of the finest of all cosmic spectacles, however: the total eclipse of the Sun. This depends on the Moon coming between us and the Sun and on it serving to block out so much of the fierce solar glare that we are able to see the delicate extensive outer atmosphere of the Sun – the corona. At present the Moon is only just able to do this: and when it has spiralled a little farther away it will not be able to produce a total eclipse at all. Conversely in earlier ages when the Moon was nearer to the Earth such eclipses must have been more frequent, more striking, and more prolonged than they now are.

W. H. Hudson

An extract from 'Hints for adder-seekers', from
The Book of a Naturalist, 1919

In spring you must go alone and softly, but you need not fear to whistle and sing, or even to shout, for the adder is deaf and cannot hear you; on the other hand, his body is sensitive in an extraordinary degree to earth vibrations, and the ordinary tread of even a very light man will disturb him at a distance of fifteen or twenty yards. That sense of the adder, which has no special organ yet may serve better than vision, hearing, smell and touch together, is of the greatest importance to it, since to a creature that lies and progresses prone on the ground and has a long brittle backbone, the heavy mammalian foot is one of the greatest dangers to its life.

Nicholas Humphrey

'The social function of intellect', chapter 2 of *Consciousness Regained*, 1984

Henry Ford, it is said, commissioned a survey of the car scrap-yards of America to find out if there were parts of the Model T Ford which never failed. His inspectors came back with reports of almost every kind of failure: axles, brakes, pistons – all were liable to go wrong. But they drew attention to one notable exception, the *kingpins* of the scrapped cars invariably had years of life left in them. With ruthless logic Ford concluded that the kingpins on the Model T were too good for their job and ordered that in future they should be made to an inferior specification.

Nature is surely at least as careful an economist as Henry Ford. It is not her habit to tolerate needless extravagance in the animals on her production lines: superfluous capacity is trimmed back, new capacity added only as and when it is needed. We do not expect to find that animals possess abilities which far exceed the calls that natural living makes on them. If someone were to argue – as I shall suggest he might argue – that some primate species (and mankind in particular) are much cleverer than they need be, we know that he is most likely to be wrong. But it is not clear why he would be wrong. This chapter explores a possible answer. It is an answer which has meant for me a rethinking of the function of intellect.

A rethinking, or merely a first-thinking? I had not previously given much thought to the biological function of intellect, and my impression is that few others have done either. In the literature on animal intelligence there has been surprisingly little discussion of how intelligence contributes to biological fitness. Comparative psychologists have established that animals of one species perform better, for instance, on the Hebb–Williams maze than those of another, or that they are quicker to pick up learning sets or more successful on an 'insight' problem; there have been attempts to relate performance on particular kinds of tests to particular underlying cognitive skills; there has (more recently) been debate on how the same skill is to be assessed with 'fairness' in animals of different species; but there has seldom been consideration given to why the animal, in its natural environment, should *need* such skill. What is the use of 'conditional oddity discrimination' to a monkey in the field? What advantage is there to an anthropoid ape in being able to recognise its own reflection in a mirror? While it might indeed be 'odd for a biologist to make it his task to explain why horses can't learn mathematics', it would not be odd for him to ask why *people can.*

The absence of discussion on these issues may reflect the view that there is little to discuss. It is tempting, certainly, to adopt a broad definition of intelligence which makes it self-evidently functional. Take, for instance, Alice Heim's definition of intelligence in man, 'the ability to grasp the essentials of a situation and respond appropriately': substitute 'adaptively' for 'appropriately' and the problem of the biological function of intellect is (tautologically) solved. But even those definitions which are not so manifestly circular tend none the less to embody value-laden words. When intelligence is defined as the 'ability' to do this or that, who dares question the biological advantage of being *able*? When reference is made to 'understanding' or 'skill at problem-solving' the terms themselves seem to quiver with adaptiveness. Every animal's world is, after all, full of things to be understood and problems to be solved. For sure, the world is full of problems – but what exactly are these problems, how do they differ from animal to animal, and what particular advantage accrues to the individual who can solve them? These are not trivial questions.

Despite what has been said, we had better have a definition of intelligence, or the discussion is at risk of going adrift. The following formula provides at least some kind of anchor: 'An animal displays intelligence when it modifies its behaviour on the basis of valid inference from evidence.' The word 'valid' is meant to imply only that the inference is logically sound; it leaves open the question of how the animal benefits in consequence. This definition is admittedly wide, since it embraces everything from simple associative learning to syllogistic reasoning. Within the spectrum it seems fair to distinguish 'low-level' from 'high-level' intelligence. It requires, for instance, relatively low-level intelligence to infer that something is likely to happen merely because similar things have happened in comparable circumstances in the past; but it requires high-level intelligence to infer that something is likely to happen because it is entailed by a *novel* conjunction of events. The former is, I suspect, a comparatively elementary skill and widespread through the animal kingdom, but the latter is much more special, a mark of the 'creative' intellect which is characteristic especially of the higher primates. In what follows I shall be enquiring into the function chiefly of 'creative' intellect.

Now I am about to set up a straw man. But he is a man whose reflection I have seen in my own mirror, and I am inclined to treat him with respect. The opinion he holds is that the main role of creative intellect lies in *practical invention*. 'Invention' here is being used broadly to mean acts of intelligent discovery by which an animal comes up with new ways of doing things. Thus it includes not only, say, the fabrication of new tools or the putting of existing objects to new use but also the discovery of new behavioural strategies, new ways of using the resources of one's own body. But, wide as its scope may be, the talk is strictly of 'practical' invention, and in this context 'practical' has a restricted meaning. For the man in question sees the need for invention as arising only in relation to the external physical environment; he has not noticed – or has not thought it important – that many animals are *social* beings.

You will see, no doubt, that I have deliberately built my straw man with feet of clay. But let us none the less see where he stands. His idea of the intellectually challenging environment has been perfectly described by Daniel Defoe. It is the desert island of Robinson Crusoe – before the arrival of Man Friday. The island is a lonely, hostile environment, full of technological challenge, a world in which Crusoe depends for his survival on his skill in gathering food, finding shelter, conserving energy, avoiding danger. And he must work fast, in a truly inventive way, for he has no time to spare for learning simply by induction from experience. But was that the kind of world in which creative intellect evolved? I believe, for reasons I shall come to, that the real world was never like that, and yet that the real world of the higher primates may in fact be considerably *more* intellectually demanding. My view – and Defoe's, as I understand him – is that it was the arrival of Man Friday on the scene which really made things difficult for Crusoe. If Monday and Tuesday, Wednesday and Thursday had turned up as well then Crusoe would have had every need to keep his wits about him.

But the case for the importance of practical invention must be taken seriously. There can be no doubt that for some species in some contexts inventiveness does seem to have survival value. The 'subsistence technology' of chimpanzees and even more that of 'natural' man involves many tricks of technique which appear prima facie to be products of creative intellect. And what is true for these anthropoids must surely be true at least in part for other species. Animals who are quick to realise new techniques (in hunting, searching, navigating, or whatever) would seem bound to gain in terms of fitness. Why, then, should one dispute that there have been selective pressures operating to bring about the evolution of intelligence in relation to practical affairs? I do not of course dispute the general principle; what I question is how much this principle *alone* explains. How clever does a man or monkey need to be before the returns on superior intellect become vanishingly small? If, despite appearances, the important practical problems of living actually demand only relatively low-level intelligence for their solution, then there would be grounds for supposing that high-level creative intelligence is wasted. Even Einstein could not get better than 100 per cent at O level. Can we really explain the evolution of the higher intellectual faculties of primates on the basis of success or failure in their 'practical exams'?

My answer is no, for the following reason: even in those species which have the most advanced technologies the exams are largely tests of knowledge rather than of imaginative reasoning. The evidence from field studies of chimpanzees all points to the fact that subsistence techniques are hardly if ever the product of premeditated invention; they are arrived at instead either by trial-and-error learning or by imitation of others. Indeed it is hard to imagine how many of the techniques could in principle be arrived at otherwise. G. Teleki concluded on the basis of his own attempts at 'termiting' that there was no way of predicting a priori what would be the most effective kind of probe to stick into a termite hill, or how best to twiddle it or, for that matter, where to stick it. He had to learn inductively by trial and error or, better, by mimicking the behaviour of Leakey, an old and experienced chimpanzee. Thus the chimpanzees' art would seem to be no more an invention than is the uncapping of milk-bottles by tits. And even where a technique could in principle be invented by deductive reasoning there are generally no grounds for supposing that it has been. Termiting by human beings is a case in point. In northern Zaire, people beat with sticks on the top of termite mounds to encourage the termites to come to the surface. The technique works because the stick-beating makes a noise like falling rain. It is just possible that someone once upon a time noticed the effect of falling rain, noticed the resemblance between the sound of rain and the beating of sticks, and put two and two together. But I doubt if that is how it happened; serendipity seems a much more likely explanation. Moreover, whatever the origin of the technique, there is certainly no reason to invoke inventiveness on the part of present-day practitioners, for these days stick-beating is culturally transmitted. My guess is that most of the practical problems that face higher primates can, as in the case of termiting, be dealt with by learned strategies without recourse to creative intelligence.

Paradoxically, I would suggest that subsistence technology, rather than requiring intelligence, may actually become a substitute for it. Provided the *social* structure of the species is such as to allow individuals to acquire subsistence techniques by simple associative learning, then there is little need for individual creativity. Thus the chimpanzees at Gombe, with their superior technological culture, may in fact have *less* need than the neighbouring baboons to be individually inventive. Indeed there might seem on the face of it to be a negative correlation between the intellectual capacity of a species and the need for intellectual output. The great apes, demonstrably the most intellectually gifted of all non-human animals, seem on the whole to lead comparatively undemanding lives, less demanding than those not only of lower primates but also of many non-primate species. During two months I spent watching gorillas in the Virunga mountains of Rwanda I could not help being struck by the fact that of all the animals in the forest the gorillas seemed to lead much the simplest existence – food abundant and easy to harvest (provided they *knew* where to find it), few if any predators (provided they *knew* how to avoid them) . . . little to do in fact (and little done) but eat, sleep and play. And the same is arguably true for natural man. Studies of contemporary Bushmen suggest that the life of hunting and gathering, typical of early man, was probably a remarkably easy one. The 'affluent savage' seems to have established a *modus vivendi* in which, for a period of perhaps five million years, he could afford to be not only physically but intellectually lazy.

We are thus faced with a conundrum. It has been repeatedly demonstrated in the artificial situations of the psychological laboratory that anthropoid apes possess impressive powers of creative reasoning, yet these feats of intelligence seem simply not to have any parallels in the behaviour of the same animals in their natural environment. I have yet to hear of any example from the field of a chimpanzee (or for that matter a Bushman) using his full capacity for inferential reasoning in the solution of a biologically relevant practical problem. Someone may retort that if an ethologist had kept watch on Einstein through a pair of field-glasses he might well have come to the conclusion that Einstein too had a humdrum mind. But that it just the point: Einstein, like the chimpanzees, displayed his genius at rare times in 'artificial' situations – he did not use it, for he did not *need* to use it, in the common world of practical affairs.

Why then do the higher primates need to be as clever as they are and, in particular, that much cleverer than other species? What – if it exists – is the natural equivalent of the laboratory test of intelligence? The answer has, I believe, been ripening on the tree of the preceding discussion. I have suggested that the life of the great apes and man may not require much in the way of practical invention, but it does depend critically on the possession of wide factual knowledge of practical technique and the nature of the habitat. Such knowledge can only be acquired in the context of a *social* community – a community which provides both a medium for the cultural transmission of information and a protective environment in which individual learning can occur. I propose that the chief role of creative intellect is to hold society together.

In what follows I shall try to explain this proposal, to justify it, and to examine some of its surprising implications.

To me, as a Cambridge-taught psychologist, the proposal is in fact a rather strange one. Experimental psychologists in Britain have tended to regard social psychology as a poor country cousin of their subject – gauche, undisciplined and slightly absurd. Let me recount how I came to a different way of thinking, since this personal history will lead directly in to what I want to say. Some years ago I made a discovery which brought home to me dramatically the fact that, even for an experimental psychologist, a *cage* is a bad place in which to keep a monkey. I was studying the recovery of vision in a rhesus monkey, Helen, from whom the visual cortex had been surgically removed. In the first four years I'd worked with her Helen had regained a considerable amount of visually guided behaviour, but she still showed no sign whatever of three-dimensional spatial vision. During all this time she had, however, been kept within the confines of a small laboratory cage. When, at length, five years after the operation, she was released from her cage and taken for walks in the open field at Madingley her sight suddenly burgeoned and within a few weeks she had recovered almost perfect spatial vision. The limits on her recovery had been imposed directly by the limited environment in which she had been living. Since that time, in working with laboratory monkeys I have been mindful of the possible damage that may have been done to them by their impoverished living conditions. I have looked anxiously through the wire mesh of the cages at Madingley, not only at my own monkeys but at Robert Hinde's. Now, Hinde's monkeys are rather better off than mine. They live in social groups of eight or nine animals in relatively large cages. But these cages are almost empty of objects, there is nothing to manipulate, nothing to explore; once a day the concrete floor is hosed down, food pellets are thrown in and that is about it. So I looked – and seeing this barren environment, thought of the stultifying effect it must have on the monkey's intellect. And then one day I looked again and saw a half-weaned infant pestering its mother, two adolescents engaged in a mock battle, an old male grooming a female while another female tried to sidle up to him, and I suddenly saw the scene with new eyes: forget about the absence of *objects*, these monkeys had *each other* to manipulate and to explore. There could be no risk of their dying an intellectual death when the social environment provided such obvious opportunity for participating in a running dialectical debate. Compared to the solitary existence of my own monkeys, the set-up in Hinde's social groups came close to resembling a simian School of Athens.

The scientific study of social interaction is not far advanced. Much of the best published literature is in fact genuinely 'literature' – Aesop and Dickens make, in their own way, as important contributions as Laing, Goffman or Argyle. But one generalisation can I think be made with certainty: the life of social animals is highly problematical. In a complex society, such as those we know exist among higher primates, there are benefits to be gained for each individual member both from preserving the overall structure of the group and at the same time from exploiting and out-manoeuvring others within it. Thus social primates are required by the very nature of the system they create and maintain to be calculating beings; they must be able to calculate the consequences of their own behaviour, to calculate the likely behaviour of others, to calculate the balance of advantage and loss – and all this in a context where the evidence on which their calculations are based is ephemeral, ambiguous and liable to change, not least as a consequence of their own actions. In such a situation, 'social skill' goes hand in hand with intellect, and here at last the intellectual faculties required are of the highest order. The game of social plot and counter-plot cannot be played merely on the basis of accumulated knowledge, any more than can a game of chess.

Like chess, a social interaction is typically a *transaction* between social partners. One animal may, for instance, wish by his own behaviour to change the behaviour of another; but since the second animal is himself reactive and intelligent the interaction soon becomes a two-way argument where each 'player' must be ready to change his tactics – and maybe his goals – as the game proceeds. Thus, over and above the cognitive skills which are required merely to perceive the current state of play (and they may be considerable), the social gamesman, like the chess-player, must be capable of a special sort of forward planning. Given that each move in the game may call forth several alternative responses from the other player this forward planning will take the form of a decision tree, having its root in the current situation and branches corresponding to the moves considered in looking ahead at different possibilities. It asks for a level of intelligence which is, I submit, unparalleled in any other sphere of living. There may be, of course, strong and weak players – yet, as master or novice, we and most other members of complex primate societies have been in this game since we were babies.

But what makes a society 'complex' in the first place? There have probably been selective pressures of two rather different kinds, one from without, the other from within society. I suggested that one of the chief functions of society is to act as it were as a polytechnic school for the teaching of subsistence technology. The social system serves the purpose in two ways: (i) by allowing a period of prolonged dependence during which young animals, spared the need to fend for themselves, are free to experiment and explore; and (ii) by bringing the young into contact with older, more experienced members of the community from whom they can learn by imitation (and perhaps, in some cases, from more formal 'lessons'). Now, to the extent that this kind of education has adaptive consequences, there will be selective pressures both to prolong the period of untrammelled infantile dependency (to increase the 'school-leaving age') and to retain older animals within the community (to increase the number of experienced 'teachers'). But the resulting mix of old and young, caretakers and dependants, sisters, cousins, aunts and grandparents not only calls for considerable social responsibility but also has potentially disruptive social consequences. The presence of dependants (young, injured or infirm) clearly calls at all times for a measure of tolerance and unselfish sharing. But in so far as biologically important resources may be scarce (as subsistence materials must sometimes be, and sexual partners will be commonly) there is a limit to which tolerance can go. Squabbles are bound to occur about access to these scarce resources and different individuals will have different interests in participating in, promoting or putting a stop to such squabbles. Thus the stage is set within the 'collegiate community' for considerable political strife. To do well for oneself while remaining within the terms of the social contract on which the fitness of the whole community ultimately depends calls for remarkable reasonableness (in both literal and colloquial senses of the word). It is no accident therefore that men, who of all primates show the longest period of dependence (nearly thirty years in the case of Bushmen), the most complex kinship structures, and the widest overlap of generations within society, should be more intelligent than chimpanzees, and chimpanzees for the same reasons more intelligent than monkeys.

Once a society has reached a certain level of complexity, then new internal pressures must arise which act to increase its complexity still further. For, in a society of the kind outlined, an animal's intellectual 'adversaries' are members of his own breeding community. If intellectual prowess is correlated with social success, and if social success means high biological fitness, then any heritable trait which increases the ability of an individual to outwit his fellows will soon spread through the gene pool. And in these circumstances there can be no going back: an evolutionary 'ratchet' has been set up, acting like a self-winding watch to increase the general intellectual standing of the species. In principle the process might be expected to continue until either the physiological mainspring of intelligence is full-wound or else intelligence itself becomes a burden. The latter seems most likely to be the limiting factor: there must surely come a point where the time required to resolve a social 'argument' becomes insupportable.

The question of the time given up to unproductive social activity is an important one. Members of the community – even if they have not evolved a runaway intellect – are bound to spend a considerable part of their lives in caretaking and social politics. It follows that they inevitably have less time to spare for basic subsistence activities. If the social system is to be of any net biological benefit the improvement in subsistence techniques which it makes possible must more than compensate for the lost time. To put the matter baldly: if an animal spends all morning in non-productive socialising, he must be at least twice as efficient a producer in the afternoon. We might therefore expect that the evolution of a social system capable of supporting advanced technology should only happen under conditions where improvements in technique can substantially increase the return on labour. This may not always be the case. To take an extreme example, the open sea is probably an environment where technical knowledge can bring little benefit, so that complex societies – and high intelligence – are contra-indicated (dolphins and whales provide, maybe, a remarkable and unexplained exception). Even at Gombe the net advantage of having a complex social system may in fact be marginal; the chimpanzees at Gombe share several of the local food resources with baboons, and it would be instructive to know how far the advantage that chimpanzees have over baboons in terms of technical skill is eroded by the relatively large amount of time they give up to social intercourse. It may be that what the chimpanzees gain on the swings of technical proficiency they lose on the roundabouts of extravagant socialising. As it is, in a year of poor harvest the chimpanzees in fact become much less sociable; my guess is that they simply cannot spare the time. The ancestors of man, however, when they moved into the savanna, discovered an environment where technical knowledge began to pay new and continuing dividends. It was in that environment that the pressures to give children an even better schooling created a social system of unprecedented complexity – and with it unprecedented challenge to intelligence.

The outcome has been the gifting of members of the human species with remarkable powers of social foresight and understanding. This social intelligence, developed initially to cope with local problems of interpersonal relationships, has in time found expression in the institutional creations of the 'savage mind' – the highly rational structures of kinship, totemism, myth and religion which characterise primitive societies. And it is, I believe, essentially the same intelligence which has created the systems of philosophical and scientific thought which have flowered in advanced civilisations in the last four thousand years. Yet civilisation has been too short-lived to have had any important evolutionary consequences; the 'environment of adaptiveness' of human intelligence remains the *social* milieu.

If man's intellect is thus suited primarily to thinking about people and their institutions, how does it fare with *non-social* problems? To end this chapter I want to raise the question of 'constraints' on human reasoning, such as might result if there is a predisposition to try to fit non-social material into a social mould.

When a person sets out to solve a social problem he may reasonably have certain expectations about what he is getting in to. First, he should know that the situation confronting him is unlikely to remain stable. Any social transaction is by its nature a developing process and the development is bound to have a degree of indeterminacy to it. Neither of the social agents involved in the transaction can be certain of the future behaviour of the other; as in Alice's game of croquet with the Queen of Hearts, both balls and hoops are always on the move. Someone embarking on such a transaction must therefore be prepared for the problem itself to alter as a consequence of his attempt to solve it – in the very act of interpretating the social world he changes it. Like Alice he may well be tempted to complain 'You've no idea how confusing it is, all the things being alive'; that is not the way the game is played at Hurlingham – and that is not the way that non-social material typically behaves. But, secondly, he should know that the development *will* have a certain logic to it. In Alice's croquet game there was real confusion, everyone played at once without waiting for turns, and there were no rules; but in a social transaction there are, if not strict rules, at least definite constraints on what is allowed and definite conventions about how a particular action by one of the transactors should be answered by the other. My earlier analogy with the chess game was perhaps a more appropriate one; in social behaviour there is a kind of turn-taking, there are limits on what actions are allowable, and at least in some circumstances there are conventional, often highly elaborated, sequences of exchange.

Even the chess analogy, however, misses a crucial feature of social interaction. For while the good chess player is essentially selfish, playing only to win, the selfishness of social animals is typically tempered by what, for want of a better term, I would call *sympathy*. By sympathy I mean a tendency on the part of one social partner to identify himself with the other and so to make the other's goals to some extent his own. The role of sympathy in the biology of social relationships has yet to be thought through in detail, but it is probable that sympathy and the 'morality' which stems from it are biologically adaptive features of the social behaviour of both men and other animals – and consequently major constraints on 'social thinking' wherever it is applied. Thus our man setting out to apply his intelligence to solve a social problem may expect to be involved in a fluid, transactional exchange with a sympathetic human partner. To the extent that the thinking appropriate to such a situation represents the customary mode of human thought, men may be expected to behave inappropriately in contexts where a transaction cannot in principle take place: if they treat inanimate entities as 'people' they are sure to make mistakes.

There are many examples of fallacious reasoning which would fit such an interpretation. The most obvious cases are those where men do in fact openly resort to animistic thinking about natural phenomena. Thus primitive – and not so primitive – peoples commonly attempt to *bargain* with nature, through prayer, through sacrifice or through ritual persuasion. In doing so they are explicitly adopting a social model, expecting nature to participate in a transaction. But nature will not transact with men; she goes her own way regardless – while her would-be interlocutors feel grateful or slighted as befits the case. Transactional thinking may not always be so openly acknowledged, but it often lies just below the surface in other cases of 'illogical' behaviour. Thus the gambler at the roulette table, who continues to bet on the red square precisely because he has already lost on red repeatedly, is behaving as though he expects the behaviour of the roulette wheel to respond eventually to his persistent overtures; he does not – as he would be wise to do – conclude that the odds are unalterably set against him. Likewise, the man in P. C. Wason's experiments on abstract reasoning who, when he is given the task of discovering a mathematical rule, typically tries to substitute *his own* rule for the predetermined one is acting as though he expects the problem itself to change in response to his trial solutions. The comment of one of Wason's subjects is revealing: 'Rules are relative. If you were the subject, and I were the experimenter, then I would be right.' In general, I would suggest, a transactional approach leads men to refuse to accept the intransigence of facts – whether the facts are physical events, mathematical axioms or scientific laws; there will always be the temptation to assume that the facts will respond like living beings to social pressures. Men expect to argue *with* problems rather than being limited to arguing *about* them.

There are times, however, when such a 'mistaken' approach to natural phenomena can be unexpectedly creative. While it may be the case that no amount of social pleading will change the weather or, for that matter, transmute base metals into gold, there are things in nature with which a kind of social intercourse is possible. It is not strictly true that nature will not transact with men. If we mean by a transaction essentially a developing relationship founded on mutual give and take, then several of the relationships which men enter into with the non-human things around them may be considered to have transactional qualities. The cultivation of plants provides a clear and interesting example: the care which a gardener gives to his plants (watering, fertilising, hoeing, pruning etc.) is attuned to the plants' emerging properties, which properties are in turn a function of the gardener's behaviour. True, plants will not respond to ordinary social pressures (though men *do* talk to them), but the way in which they give to and receive from a gardener bears, I suggest, a close structural similarity to a simple social relationship. If C. Trevarthen can speak of 'conversations' between a mother and a two-month-old baby, so too might we speak of a conversation between a gardener and his roses or a farmer and his corn. And the same can be argued for men's interactions with certain wholly inanimate materials. The relationship of a potter to his clay, a smelter to his ore or a cook to his soup are all relationships of fluid mutual exchange, again proto-social in character.

It is not just that transactional thinking is typical of man; transactions are something which people actively seek out and will force on nature wherever they are able. In the Doll Museum in Edinburgh there is a case full of bones clothed in scraps of rag – moving reminders of the desire of human children to conjure up social relationships with even the most unpromising material. Through a long history, men have, I believe, explored the transactional possibilities of countless of the things in their environment, and sometimes, Pygmalion-like, the things have come alive. Thus many of mankind's most prized technological discoveries, from agriculture to chemistry, may have had their origin not in the deliberate application of practical intelligence but in the fortunate misapplication of social intelligence.

The rise of classical scientific method has in large measure depended on human thinkers disciplining themselves to abjure transactional, socio-magical styles of reasoning. But scientific method has come to the fore only in the last few hundred years of mankind's history, and in our own times there are everywhere signs of a return to more magical systems of interpretation. In dealing with the non-social world the former method is undoubtedly the more immediately appropriate; but the latter is perhaps more natural to man. Transactional thinking may indeed be irrepressible: within the most disciplined Jekyll is concealed a transactional Hyde. Charles Dodgson the mathematician shared his pen amicably enough with Lewis Carroll the inventor of Wonderland, but the split is often neither so comfortable nor so complete. Newton is revealed in his private papers as a Rosicrucian mystic, and his intellectual descendants continue to this day to apply strange double standards to their thinking – witness the way in which certain British physicists took up the cause of Uri Geller, the man who, by wishing it, could bend a metal spoon. In the long view of science, there is, I suspect, good reason to approve this kind of inconsistency. For while 'normal science' (in Kuhn's sense of the term) has little if any room for social thinking, 'revolutionary science' may more often than we realise derive its inspiration from a vision of a socially transacting universe.

Particle physics has already followed Alice down the rabbit-hole into a world peopled by 'families' of elementary particles endowed with 'strangeness' and 'charm'. *Vide*, for example, the following report in *New Scientist*: 'The particles searched for at SPEAR were the *cousins* of the psis made from one *charm* quark and one *uncharmed* antiquark. This contrasts with the *siblings* of the psis . . .' (my italics). Who knows where such 'socio-physics' may eventually lead?

The ideology of classical science has had a huge but in many ways narrowing influence on ideas about the nature of 'intelligent' behaviour. But no matter what the high priests, from Bacon to Popper, have had to say about how people ought to think, they have never come near to describing how people *do* think. In so far as an idealised view of scientific method has been the dominant influence on mankind's recent intellectual history, biologists should be the first to follow Henry Ford in dismissing recent history as 'bunk'. Evolutionary history, however, is a different matter. The formative years for human intellect were the years when man lived as a social savage on the plains of Africa. Even now, as Sir Thomas Browne wrote in *Religio medici*, 'All Africa and her prodigies are within us.'

Julian Huxley

An extract from *New Bottles for Old Wine*, 1959

Evolution on the human level, although it has been operating for the barest fraction of geological time, has already produced very extraordinary new results, impossible even to conceive of on the biological level – for example, Dante's *Divina Commedia*, guided missiles, Picasso's *Guernica*, Einstein's theory of relativity, ritual cannibalism, the Parthenon, the Roman Catholic Church, the films of the Marx Brothers, modern textile mills, Belsen, and the mystical experiences of Buddhist saints. Most extraordinary in principle, it has generated values.

T. H. Huxley

'On a piece of chalk', British Association lecture, 1868

If a well were sunk at our feet in the midst of the city of Norwich, the diggers would very soon find themselves at work in that white substance almost too soft to be called rock, with which we are familiar as 'chalk'.

Were the thin soil which covers it all washed away, a curved band of white chalk might be followed diagonally across England from Lulworth in Dorset to Flamborough Head in Yorkshire – over 280 miles as the crow flies. From this band to the North Sea, on the east, and the Channel, on the south, the chalk is largely hidden by other deposits; but, except in the Weald of Kent and Sussex, it enters into the very foundation of all the south-eastern counties.

What is this wide-spread component of the surface of the earth, and whence did it come?

You may think this no very helpful inquiry. You may not unnaturally suppose that the attempt to solve such problems as these can lead to no result, save that of entangling the inquirer in vague speculations, incapable of refutation and of verification. If such were really the case, I should have selected some other subject for my discourse. But, in truth, after much deliberation, I have been unable to think of any topic which would so well enable me to lead you to see how solid is the foundation upon which some of the most startling conclusions of physical science rest.

A great chapter of the history of the world is written in the chalk.

To the unassisted eye, chalk looks like a very loose and open kind of stone. But it is possible to grind a slice of chalk down so thin that you can see through it – until it is thin enough, in fact, to be examined under the microscope. The general mass of it is made up of very minute granules; but, embedded in this matrix, are innumerable bodies, each of which may be proved to be a beautifully-constructed calcareous fabric, made up of a number of chambers, communicating freely with one another. The chambered bodies are of various forms. One of the commonest is something like a badly-grown raspberry, being formed of a number of nearly globular chambers of different sizes congregated together. It is called *Globigerina*. Globigerinae are exclusively marine animals, the skeletons of which abound at the bottom of deep seas: there is no escaping the conclusion that the chalk itself is the dried mud of an ancient deep sea.

I was surprised to find that many of what I have called the 'granules' of that mud were not the mere powder and waste of the globigeriniae, but that they had a definite form and size. I termed these bodies 'coccoliths'. Not unfrequently, bodies similar to these 'coccoliths' were aggregated together into spheroids, which I termed 'coccospheres'. So far as we knew, these bodies were peculiar to the Atlantic. But a few years ago, Mr Sorby, in making a careful examination of the chalk by means of thin sections and comparing these formed particles with those in the Atlantic, he found the two to be identical. Here was a further and most interesting confirmation, from internal evidence, of the essential identity of the chalk with modern deep-sea mud. Globigeriniae, coccoliths, and coccospheres are the chief constituents of both.

I think you will now allow that I did not overstate my case when I asserted that we have as strong grounds for believing that all the vast area of dry land, at present occupied by the chalk, was once at the bottom of the sea, as we have for any matter of history whatever; while there is no justification for any other belief.

The population of the world had undergone slow and gradual, but incessant changes. One species has vanished and another has taken its place, as time has passed on. It is by the population of the chalk sea that the ancient and modern inhabitants of the world are most completely connected. Thus the chalk contains remains of those strange flying and swimming reptiles, the pterodactyl, the ichthyosaurus, and the plesiosaurus, which are found in no later deposits, but abounded in preceding ages.

Up to this moment I have stated, so far as I know, nothing but well-authenticated facts. But the mind is so constituted that it does not willingly rest in facts and immediate causes, but seeks always after a knowledge of the remoter links in the chain of causation.

There is not a shadow of a reason for believing that the physical changes of the globe, in past times, have been affected by other than natural causes. Is there any more reason for believing that the concomitant modifications in the forms of the living inhabitants of the globe have been brought about in other ways? Science gives no countenance to such a wild fancy; nor can even the perverse ingenuity of a commentator pretend to discover this sense, in the simple words in which the writer of Genesis records the proceedings of the fifth and sixth days of the Creation.

A small beginning has led us to a great ending. If I were to put the bit of chalk with which we started into the hot but obscure flames of burning hydrogen, it would presently shine like the sun. It seems to me that this physical metamorphosis is no false image of what has been the result of our subjecting it to a jet of fervent, though nowise brilliant, thought tonight. It has become luminous.

Arthur Koestler

'Cosmic limerick'

Young Archie, the intrepid mole,
Went down to explore a Black Hole.
  A stark singularity,
  Devoid of all charity,
Devoured the mole as a whole.

Primo Levi

'Travels with C', from *The Periodic Table*, 1984

What chemist, facing the elements of the periodic table, or scanning the monumental indices of Beilstein or Landolt, does not recognize scattered among them the tatters or trophies of his own professional past? He has only to leaf through any treatise for memories to assail him: somewhere there is a colleague among us who has tied his destiny, indelibly, to bromide or propylene, or to the NCO group or glutamic acid. Every chemistry student, faced by almost any treatise, should be aware that his future is also written in indecipherable characters on one of its pages, perhaps in a single line, formula, or word. Every no-longer-young chemist, turning again to the *verängnisvoll* page in that same treatise, is struck by love or disgust, delight or despair.

So it happens that every chemical element says something different to each of us, as do the mountain valleys or beaches visited in youth. One might make an exception for carbon, because it says everything to everyone: it is not specific, in the same way that Adam is not specific as an ancestor. Carbon is the only element that can bind itself in long, stable chains without a great expense of energy, and for life on earth long chains are required. Therefore carbon is the key element of living substance, but its entry into the living world is not easy and must follow an obligatory and intricate path. If the creating of organic compounds were not a common daily occurrence, on the scale of billions of tons a week, wherever the green of a leaf appears, it would by full right deserve to be called a miracle.

It was to carbon, the element of life, that my first literary dream was turned – and now I want to tell the story of a single atom of carbon.

My fictional character lies, for hundreds of millions of years, bound to three atoms of oxygen and one of calcium, in the form of limestone. (It already has behind it a very long cosmic history, but that we shall ignore.) Time does not exist for it, or exists only in the form of sluggish daily or seasonal variations in temperature. Its existence, whose monotony cannot be conceived of without horror, is an alternation of hots and colds.

The limestone ledge of which the atom forms a part lies within reach of man and his pickax. At any moment – which I, as narrator, decide out of pure caprice to be the year 1840 – a blow of the pickax detached the limestone and sent it on its way to the lime furnace, where it was plunged into the world of things that change. The atom of carbon was roasted until it separated from the limestone's calcium, which remained, so to speak, with its feet on the ground and went on to meet a less brilliant destiny. Still clinging firmly to two of its three former oxygen companions, our fictional character issued from the chimney and rode the path of the air. Its story, which once was immobile, now took wing.

The atom was caught by the wind, flung down onto the earth, lifted ten kilometers high. It was breathed in by a falcon, but did not penetrate the bird's rich blood and was exhaled. It dissolved three times in the sea, once in the water of a cascading torrent, and again was expelled. It traveled with the wind for eight years – now high, now low, on the sea and among the clouds, over forests, deserts, and limitless expanses of ice. Finally, it stumbled into capture and the organic adventure.

The year was 1848. The atom of carbon, accompanied by its two satellites of oxygen, which maintained it in a gaseous state, was borne by the wind along a row of vines. It had the good fortune to brush against a leaf, penetrate it, and be nailed there by a ray of the sun. On entering the leaf, it collided with other innumerable molecules of nitrogen and oxygen. It adhered to a large and complicated molecule that activated it, and simultaneously it received the decisive message from the sky, in the flashing form of a packet of solar light: in an instant, like an insect caught by a spider, the carbon atom was separated from its oxygen, combined with hydrogen, and finally inserted in a chain of life. All this happened swiftly, in silence, at the temperature and pressure of the atmosphere.

Now our atom was part of a structure, in an architectural sense, and had become tied to five companions so identical with it that only my fictional narrative permits me to distinguish among them. It was a beautiful ring-shaped structure, an almost even-sided hexagon, which was subject to complicated exchanges and balances with the water in which it was dissolved – because by now the atom was dissolved in water, indeed in the lymph of life, and to remain dissolved is both the obligation and the privilege of all substances that are destined to change.

The atom had entered the leaf to form part of a molecule of glucose, an intermediary phase that prepared it for its first contact with the animal world but did not authorize it to take on the higher responsibility of becoming part of a proteinic edifice. Hence it traveled, at the slow pace of vegetal juices, from the leaf to the petiole through the stem to the pedicel, where it descended to the nearly ripe bunch of grapes. What then followed is the province of the wine makers: we are interested only in pointing out that it escaped alcoholic fermentation (to our advantage, since we would not know how to put it in words) and reached the wine without changing its nature.

The destiny of wine is to be drunk, and it is the destiny of glucose to be oxidized. But it was not oxidized immediately: its drinker kept it in his liver for more than a week, well curled up and tranquil, as reserve aliment for a sudden effort – an effort he was forced to make the following Sunday, pursuing a bolting horse.

Farewell to the hexagonal structure. In the space of a few instants, the skein was unwound and became glucose again, and this was dragged by the current in the bloodstream all the way to a minute fiber in the thigh, and there was brutally split into two molecules of lactic acid – the grim harbinger of fatigue. Only some minutes later were the lungs able to supply the oxygen necessary to oxidize the acid quietly. So a new molecule of carbon dioxide returned to the atmosphere, and a parcel of the energy that the sun handed to the vine-shoot passed from the state of chemical energy to that of mechanical energy and so settled down in the slothful condition of heat, warming up imperceptibly the air moved by the running and blood of the runner.

Our atom was again carbon dioxide. The wind this time carried it over the Apennines and the Adriatic, Greece, the Aegean, and Cyprus; over Lebanon, the dance was repeated. The atom became trapped in a structure that promised to last for a long time: the venerable trunk of a cedar. It passed again through the stages we have already described, and the glucose of which it was a part belonged, like the bead of a rosary, to a long chain of cellulose. This was no longer the geologic fixity of limestone, measured in millions of years, but we can easily speak of centuries because the cedar is a tree of great longevity.

The year 1968 began the next episode in the life of our fictional character. A woodworm had taken an interest in the cedar. It had dug its tunnel between the wood and the bark, with the obstinate and blind voracity of its kind; as it drilled it grew, and its tunnel grew with it. It became a pupa, and in the spring it emerged in the shape of an ugly gray butterfly that was drying in the open world, confused and dazzled by the splendor of the sun.

Our atom was in one of the insect's thousand eyes, contributing to the crude and summary vision with which the butterfly orients itself in space. The insect was fecundated, laid its eggs, and died. The small cadaver lay in the undergrowth of the woods, emptied of its fluids, but the chitinous carapace, almost indestructible, resisted decay for a long time. The snow and sun did not injure it: it was buried by the dead leaves and the loam – it had become a slough, a 'thing'.

Still, the death of atoms, unlike ours, is never irrevocable. Here at work were the omnipresent, untiring, and invisible gravediggers of the undergrowth, the microorganisms of the humus. The carapace, with its eyes now blind, had slowly disintegrated, and the ex-drinker, ex-cedar, ex-woodworm, had once again taken wing.

It is possible to show that my completely arbitrary narrative is true. I could tell innumerable other stories, and they would all be literally true, in the nature of the transitions, in their order and data. The number of atoms is so great that one could always be found whose story coincided with the most whimsically invented tale. I could recount an endless number of narratives about carbon atoms that became colors or perfumes in flowers; of others that, from tiny algae to small crustaceans to fish, gradually returned carbon monoxide to the waters of the sea in a perpetual round dance of life and death, in which every devourer is immediately devoured; of others that attained a decorous semi-eternity in the yellowed pages of some archival document, or on the canvas of a famous painter; of those to which fell the privilege of forming part of a grain of pollen and that left their fossil imprint in the rocks for our curiosity; of others still that descended to become part of the mysterious messengers of the shape of the human seed and participated in the subtle process of division, duplication, and fusion from which each of us is born. Instead, I will tell the story of only one more, the most secret, and I will tell it with the humility and constraint of him who knows from the start that the trade of clothing facts in words is bound by its very nature to fail.

Our atom of carbon is again among us, in a glass of milk. It is inserted in a very complex, long chemical chain, yet such that almost all of its links are acceptable to the human body. It is then swallowed. Since every living structure harbors a savage distrust toward every contribution of diverse material of living origin, the chain is meticulously shattered and its fragments, one by one, are accepted or rejected. One – the one that concerns us – crosses the intestinal threshold and enters the bloodstream, where the atom migrates, knocks at the door of a nerve cell, enters, and supplants the carbon that was part of it. This cell belongs to a brain, and it is my brain; the cell in question, and within it the atom in question, are in charge of my writing, in a mysterious game that nobody has yet described. It is that which at this instant, issuing out of a labyrinthine tangle of yes and no, makes my hand run along a certain path on the paper, mark it with these volutes that are signs: a double snap, up and down, between two levels of energy, guides this hand of mine to impress on this paper this dot here, *this* one.

Alan Lightman

'Smile' from *Science 85*

The woman's lips are glistening in the sunlight, reflecting high-density light onto the back of the man's retina.

It is an afternoon in March. A man and a woman stand on the wooden dock, gazing at the lake and the waves on the water. They haven't noticed each other.

The man turns. And so begins the sequence of biochemical events informing him of her. Light reflected from her body instantly enters the pupils of his eyes, at the rate of 10 trillion particles of light per second. Once through the pupil of each eye, the light travels through an oval-shaped lens, then through a transparent, jellylike substance filling up the eyeball, and lands on the retina. Here it is gathered by a hundred million rod and cone cells.

Cells in the path of reflected highlights receive a great deal of light; cells falling in the shadows of the reflected scene receive very little. The woman's lips, for example, are just now glistening in the sunlight, reflecting light of high intensity onto a tiny patch of cells slightly north-east of back center of the man's retina. The edges around her mouth, on the other hand, are rather dark, so that cells neighboring the north-east patch receive much less light.

Each particle of light ends its journey to the eye upon meeting a retinene molecule, consisting of 20 carbon atoms, 28 hydrogen atoms, and one oxygen atom. In its dormant condition, each retinene molecule is attached to a protein molecule and has a twist between the 11th and 15th carbon atoms. But when light strikes it, as is now happening in about 30,000 trillion retinene molecules every second, the molecule straightens out and separates from its protein. After several intermediate steps, it wraps into a twist again, awaiting arrival of a new particle of light. Far less than a thousandth of a second has elapsed since this man saw that woman.

Triggered by the dance of the retinene molecules, the nerve cells, or neurons, respond. First in the eye and then in the brain. One neuron, for instance, has just gone into action. Protein molecules on its surface suddenly change their shape, blocking the flow of positively charged sodium atoms from the surrounding body fluid. This change in flow of electrically charged atoms produces a change in voltage that shudders through the cell. After a distance of a fraction of an inch, the electrical signal reaches the end of the neuron, altering the release of specific molecules, which migrate a distance of a hundred-thousandth of an inch until they reach the next neuron, passing along the news.

The woman, in fact, holds her hands by her sides and tilts her head at an angle of five and a half degrees. Her hair falls just to her shoulders. This information and much much more is exactingly encoded by the electrical pulses in the various neurons of the man's eyes.

In another few thousandths of a second, the electrical signals reach the ganglion neurons, which bunch together in the optic nerve at the back of the eye and carry their data to the brain. Here the impulses race to the primary visual cortex, a highly folded layer of tissue about a 10th of an inch thick and two square inches in area, containing 100 million neurons in half-a-dozen layers. The fourth layer receives the input first, does a preliminary analysis, and transfers the information to neurons in other layers. At every stage, each neuron may receive signals from a thousand other neurons, combine the signals – some of which cancel each other out – and dispatch the computed result to a thousand-odd other neurons.

After about 30 seconds – after several hundred trillion particles of reflected light have entered the man's eyes and been processed – the woman says hello. Immediately, molecules of air are pushed together, then apart, then together, beginning in her vocal chords and travelling in a springlike motion to the man's ears. The sound makes the trip from her to him (20 feet) in a 50th of a second.

Within each of his ears, the vibrating air quickly covers the distance to the eardrum. The eardrum, an oval membrane about 0.3 inch in diameter and tilted 55 degrees from the floor of the auditory canal, itself begins trembling and transmits its motion to three tiny bones. From there, the vibrations shake the fluid in the cochlea, which spirals snail-like two-and-a-half turns around.

Inside the cochlea the tones are deciphered. Here, a very thin membrane undulates in step with the sloshing fluid, and through this basilar membrane run tiny filaments of varying thicknesses, like strings on a harp. The woman's voice, from afar, is playing this harp. Her hello begins in the low registers and rises in pitch toward the end. In precise response, the thick filaments in the basilar membrane vibrate first, followed by the thinner ones. Finally, tens of thousands of rod-shaped bodies perched on the basilar membrane convey their particular quiverings to the auditory nerve.

News of the woman's hello, in electrical form, races along the neurons of the auditory nerve and enters the man's brain through the thalamus to a specialized region of the cerebral cortex for further processing.

Eventually, a large fraction of the trillion neurons in the man's brain become involved with computing the visual and auditory data just acquired. Sodium and potassium gates open and close. Electrical currents speed along neuron fibers. Molecules flow from one nerve ending to the next.

All of this is known. What is not known is why, after about a minute, the man walks over to the woman and smiles.

Percival Lowell

An extract from the Conclusion to Mars and its Canals, 1906

That Mars is inhabited by beings of some sort or other we may consider as certain as it is uncertain what those beings may be. The theory of the existence of intelligent life on Mars may be likened to the atomic theory in chemistry in that in both we are led to the belief in units which we are alike unable to define. Both theories explain the facts in their respective fields and are the only theories that do, while as to what an atom may resemble we know as little as what a Martian may be like. But the behavior of chemic compounds points to the existence of atoms too small for us to see, and in the same way the aspect and behavior of the Martian markings implies the action of agents too far away to be made out.

But though in neither case can we tell anything of the bodily form of its unit, we can in both predicate a good deal about their workings. Apart from the general fact of intelligence implied by the geometric character of their constructions, is the evidence as to its degree afforded by the cosmopolitan extent of the action. Girdling their globe and stretching from pole to pole, the Martian canal system not only embraces their whole world, but is an organized entity. Each canal joins another, which in turn connects with a third, and so on over the entire surface of the planet. This continuity of construction posits a community of interest. Now, when we consider that though not so large as the Earth the world of Mars is one of 4200 miles diameter and therefore containing something like 212,000,000 of square miles, the unity of the process acquires considerable significance. The supposed vast enterprises of the earth look small beside it. None of them but become local in comparison, gigantic as they seem to us to be.

The first thing that is forced on us in conclusion is the necessarily intelligent and non-bellicose character of the community which could thus act as a unit throughout its globe. War is a survival among us from savage times and affects now chiefly the boyish and unthinking element of the nation. The wisest realize that there are better ways for practicing heroism and other and more certain ends of insuring the survival of the fittest. It is something a people outgrow. But whether they consciously practice peace or not, nature in its evolution eventually practices it for them, and after enough of the inhabitants of a globe have killed each other off, the remainder must find it more advantageous to work together for the common good. Whether increasing common sense or increasing necessity was the spur that drove the Martians to this eminently sagacious state we cannot say, but it is certain that reached it they have, and equally certain that if they had not they must all die. When a planet has attained to the age of advancing decrepitude, and the remnant of its water supply resides simply in its polar caps, these can only be effectively tapped for the benefit of the inhabitants when arctic and equatorial peoples are at one. Difference of policy on the question of the all-important water supply means nothing short of death. Isolated communities cannot there be sufficient unto themselves; they must combine to solidarity or perish. . . .

Gwyn Macfarlane

An extract from *Howard Florey*, 1979

. . . Scientists are, in general, not very different as people from non-scientists; they display the same variety of personalities, emotions, ambitions, and interests that one encounters in most other professions. And not all scientists are dedicated to research: for many, science means routine laboratory work or teaching. But scientists have in common a disciplined scepticism that demands concrete proof for any statement or belief concerning the material world. This is the legacy of the 'scientific method' – a phrase seldom used by scientists themselves, and a method practised by them instinctively rather than consciously. To be a great scientist needs the same sort of refinement and power that raises an artist, an engineer, or a statesman to the highest levels of achievement. Some scientists have become great men in the academic world, in government, or in public life without being great scientists. To be a great scientist one must (among other things) be a pioneer, a discoverer, a dedicated research worker, and one must be successful.

[Perhaps predominant among t]he qualities that made Florey . . . [a] great scientist was that most basic one: objective honesty. Self-delusion – the unconscious misreading or selection of results to support some cherished hope or carefully constructed theory and a disregard for misfit facts – this is the greatest enemy of true progress. The literature of science is lumbered with theories – the beloved brainchildren of their creators – that are little more than dogma. Florey was quick to see those that could easily be demolished by experiment, and much of his earlier work consisted of such clearances. He was equally objective about his own researches, distrusting any ideas that could not be put to the test at once, and distrusting laboratory results until they were confirmed again and again. His characteristic attitude of scepticism and laconic understatement, so often and rather dauntingly applied to the work of others, was applied no less rigidly to his own work. But this was the negative side of his approach. On the positive side, he could see more quickly than most the really significant fact in a jumble of observations and pretentious ideas, seize upon it, and develop it.

A most important factor in Florey's practical success as a scientist was his sense of direction in research. His flair for choosing lines that led, not into blind alleys, but into wider and wider fields has been described as 'almost uncanny'. The studies of the micro-circulation, inflammation, vascular pathology, the lymphatic system, the functions of the lymphocyte, mucus secretion, gastro-intestinal function, and lysozyme all preceded his work on penicillin and almost all continued to yield important results throughout the forty years of his active research life. But 'uncanny' is not a word favoured by scientists, and in this connection it conveys the idea of a mysteriously predictive intuition that most would reject. Chance, of course, plays a very large part in the success or failure of scientific research; and once more it must be stressed that the apparently most fortunate scientist is often, in reality, the most critically observant. Good luck undoubtedly made some contribution to Florey's long list of successful research projects, but it is stretching the bounds of probability too far to suppose that he was merely lucky in happening to choose, among the countless attractive openings that are revealed at every step along any path in research, just those that would lead on and out into wider fields. It is easier to believe that he did possess an instinctive gift, recognizing pointers and indicators not obvious to less gifted people. We admit that other such gifts as artistic inspiration, musical or mathematical genius, or even the homing abilities of birds and animals, are at present inexplicable by the experimental psychologist or physiologist. Perhaps a sense of direction in research is of the same order and neither more nor less mysterious.

The importance of logic in research depends on the sort of work being done. Where the number of variables is small and the basic rules ('laws') are well established (as in some branches of physics or chemistry) the significance of a new observation can be logically deduced with fair confidence. But in biology even the number of variables is uncertain and the rules themselves have often to be discovered as the research proceeds. Deductive logic is thus unreliable and often comes to depend on statistical analysis and unsatisfactory shades of probability. The best biologist is not necessarily, therefore, the best logician, and we cannot entrust his research to a computer. The other aspect of the logical approach – induction – is not a matter of fixed rules and pure reason (despite the canons of John Stuart Mill and the faith of Almroth Wright). Induction requires an effort of the imagination, the creation of a picture of the world that *might* (not must) be true, and which can only be verified by experiment. Florey was essentially an experimentalist. 'If you do the experiment you may not be certain to get an answer,' Florey used to say, 'but if you don't do it you can be certain not to get one.' 'Do the experiment' was his motto, as it had been for John Hunter. From the results of previous work he would think out the experiment most likely to give useful results, and design it with care to combine the greatest economy of time, effort, and materials with the maximum of reliable information. He seldom took more than a single inductive step without verification, and he relied on multiple experiments to offset the inevitable variability of any biological result. But he always tried to achieve simple, clear-cut answers by breaking down a problem into simple clear-cut questions, and he disliked and distrusted experiments so variable that statistics were needed to evaluate their results.

Florey's success was also due to a prodigious industry coupled with supreme technical skill, and a concentration that would allow of no interruption. During the early years it was his habit to do an experiment every day, seven days a week, except for the annual holidays that were often partly spent in foreign laboratories. An experiment for Florey was not a matter of some test-tubes and reagents, or of a few sections and a microscope; it was usually a matter of animal surgery involving hours of operating and the most meticulous and exacting techniques. There are hundreds of such experiments carefully recorded in his notebooks. He acquired, thereby, a mass of experimental data and his published papers were based so soundly that he could be certain of his results and of the validity of conclusions that never went beyond the facts. This style of research is in contrast to the evanescent excursions of workers like Fleming on the one hand, and the profound originality of those like Sherrington on the other. Looking back over Florey's papers one seldom finds a major line of research that began with an original idea of his own. Some of his early work was suggested by Sherrington, and some by his recognition of fallacies in the published papers of other people. The idea of working on mucus secretion, however, prompted by his own gastritis, seems to have occurred to him *de novo*. But if Florey cannot be rated as a brilliant originator of ideas, he was the best and soundest builder of solid factual knowledge in British experimental pathology.

One of Florey's characteristics was his capacity to attract collaborators. He worked with them because of their ability and not necessarily because he liked them. 'I would work with the devil himself – if he were good enough,' he used to say. Though, as his letters have shown, he seems to have thought of himself as somehow isolated from his fellow-men, he never lacked people eager to work with him. Some stayed with him, or in his department, for the greater part of their lives, and many of his postgraduate students came back to him for further working visits whenever they could. At first he worked with one or possibly two collaborators on any one problem; later the idea of the team took shape. Lysozyme occupied five people, including Florey himself, though the association was a loose one. The lymphocyte project, involving eight graduates, was planned in every detail by Florey, and the penicillin work, of course, was a sustained and co-ordinated effort by a still larger group. What was the attraction that brought these people to Florey's laboratory and kept them there to do, in many cases, the best work of their careers with him? It was not the money or the security that he could offer them, since he had little to give of either. It was not the promise of an easy job: Florey bluntly promised them hard work, no certainty of success, and no future in his department if they failed. Nor was it any particular charm of manner: Florey until quite late in life was brusque, sometimes brutally direct, seldom complimentary, and never flattering. Nevertheless, it was Florey's personality that captured so many able young research workers, who had been initially attracted by his scientific reputation. What they found in him was an infectious vitality: great physical energy combined with an independence of mind that seemed to open mental windows and let the fresh air of realism clear away stuffy academic pomposities. Above all they gained something of his own attitude to research, the sense of purpose, the excitement of discovery (which he seldom expressed in words), and the will to work. And he gave them confidence. Despite his show of disparagement and pessimism, they felt that, like a good ship's captain, he was always in control, and knew exactly where he and they were going. Like a good captain, too, he treated his crew fairly and well. He knew (and cared) far more about their personal difficulties and troubles than most of them ever realized, and his offhand manner concealed an unsuspected sensitivity.

The main driving force in Florey's life was his pleasure in experimental research. This, in turn, had two sources. There was, first, the excitement of possible or actual discovery, the lure that is felt by every explorer or experimentalist. Florey, in his letters, expressed the thrill, almost the sense of awe, that he experienced when he realized that he was probably the first person ever to see some natural phenomenon. This involves the sense of priority which is often misunderstood or underestimated by non-explorers. 'Priority is a word that figures prominently in the thoughts and vocabulary of most contemporary scientists,' wrote Florey in 1963. 'Like geographical explorers of old, the scientist likes to be the first to make a discovery, the first to do something.' The second factor in Florey's enjoyment of research was in the design and the technical performance of experiments. He loved overcoming natural difficulties, just as he had enjoyed defeating human adversaries in any form of competition or confrontation. He enjoyed practical planning, but his greatest pleasure was in putting his plans into effect in the laboratory. He preferred research ideas to be simple, but he was quite prepared to pursue them by elaborate and very difficult methods. He was a superb technician and the exercise of his skill gave him the same sort of satisfaction as that experienced by a master craftsman. With the thoroughness that had marked all his work from his schooldays onwards, he had prepared himself to acquire these techniques by studying with their established masters in Europe and America. And once he had perfected a technique he would apply it with tenacity. Florey, as Drury had remarked, was 'a great finisher'. Fleming was like a man who stumbles on a nugget of gold, shows it to a few friends, and then goes off to look for something else. Florey was like a man who goes back to the same spot and creates a gold mine.

Without the evidence of his letters to Ethel, it would be easy to be misled by the picture of his own character that he was at pains to present. Though socially friendly, and certainly amusing in a somewhat astringent way, Florey almost invariably kept his personal relations, even with people he had known and worked with for years, on a superficial level. Except for members of his own family and John Fulton, he did not use Christian names. Since in his professional life he seldom displayed obvious enthusiasm or optimism, the source of his drive and energy was mysterious for most of his colleagues. But we have seen a different character in the writer of the letters, and it is impossible to believe that this character was not still at the heart of his personality, however carefully concealed. The Florey who wrote to Ethel Reed was a sensitive young man, desperately anxious to succeed in research, deeply appreciative of the arts, but so obviously unsure of himself that, even then, one has to perceive his true feelings through a protective screen of understatement. He was lonely and craved the sympathetic companionship that he hoped to attain through marriage. When this failed to materialize he retreated even further into his emotional shell, and his need to justify himself had to be satisfied more and more through his work. Even here he feared disappointments, and he guarded himself against possibly painful experiences by a studied under-valuation of the chances. He kept potential friends at arm's length, where they would be unlikely to disappoint or hurt him, but he had a very real concern for people. He maintained publicly that the idea of 'suffering humanity' was not in his mind when he started work on penicillin, but it had been very much in his mind fifteen years before when he wrote (about a possible treatment for tuberculosis): 'One becomes rather lost in a maze at the thought of stopping the appalling thing of seeing young people maimed and wiped out while one can do nothing.' This emotional reaction surely reveals one of the hidden motives behind Florey's work. If so, few humane intentions have been more abundantly fulfilled.

James Clerk Maxwell

Extracts from his *Scientific Papers*, 1890/1891

The human mind is seldom satisfied, and is certainly never exercising its highest functions, when it is doing the work of a calculating machine. What the man of science, whether he is a mathematician or a physical inquirer, aims at is, to acquire and develope clear ideas of the things he deals with. For this purpose he is willing to enter on long calculations, and to be for a season a calculating machine, if he can only at last make his ideas clearer.

But if he finds that clear ideas are not to be obtained by means of processes the steps of which he is sure to forget before he has reached the conclusion, it is much better that he should turn to another method, and try to understand the subject by means of well-chosen illustrations derived from subjects with which he is more familiar.

We all know how much more popular the illustrative method of exposition is found, than that in which bare processes of reasoning and calculation form the principal subject of discourse.

Now a truly scientific illustration is a method to enable the mind to grasp some conception or law in one branch of science, by placing before it a conception or a law in a different branch of science, and directing the mind to lay hold of that mathematical form which is common to the corresponding ideas in the two sciences, leaving out of account for the present the difference between the physical nature of the real phenomena.

The correctness of such an illustration depends on whether the two systems of ideas which are compared together are really analogous in form, or whether, in other words, the corresponding physical quantities really belong to the same mathematical class. When this condition is fulfilled, the illustration is not only convenient for teaching science in a pleasant and easy manner, but the recognition of the formal analogy between the two systems of ideas leads to a knowledge of both, more profound than could be obtained by studying each system separately.

There are men who, when any relation or law, however complex, is put before them in a symbolical form, can grasp its full meaning as a relation among abstract quantities. Such men sometimes treat with indifference the further statement that quantities actually exist in nature which fulfil this relation. The mental image of the concrete reality seems rather to disturb than to assist their contemplations.

But the great majority of mankind are utterly unable, without long training, to retain in their minds the unembodied symbols of the pure mathematician, so that, if science is ever to become popular, and yet remain scientific, it must be by a profound study and a copious application of those principles of the mathematical classification of quantities which, as we have seen, lie at the root of every truly scientific illustration.

There are, as I have said, some minds which can go on contemplating with satisfaction pure quantities presented to the eye by symbols, and to the mind in a form which none but mathematicians can conceive.

There are others who feel more enjoyment in following geometrical forms, which they draw on paper, or build up in the empty space before them.

Others, again, are not content unless they can project their whole physical energies into the scene which they conjure up. They learn at what a rate the planets rush through space, and they experience a delightful feeling of exhilaration. They calculate the forces with which the heavenly bodies pull at one another, and they feel their own muscles straining with the effort.

To such men momentum, energy, mass are not mere abstract expressions of the results of scientific inquiry. They are words of power, which stir their souls like the memories of childhood.

For the sake of persons of these different types, scientific truth should be presented in different forms, and should be regarded as equally scientific, whether it appears in the robust form and the vivid colouring of a physical illustration, or in the tenuity and paleness of a symbolical expression. [1870]

This characteristic of modern experiments – that they consist principally of measurements, – is so prominent, that the opinion seems to have got abroad, that in a few years all the great physical constants will have been approximately estimated, and that the only occupation which will then be left to men of science will be to carry on these measurements to another place of decimals.

If this is really the state of things to which we are approaching, our Laboratory may perhaps become celebrated as a place of conscientious labour and consummate skill, but it will be out of place in the University, and ought rather to be classed with the other great workshops of our country, where equal ability is directed to more useful ends.

But we have no right to think thus of the unsearchable riches of creation, or of the untried fertility of those fresh minds into which these riches will continue to be poured. It may possibly be true that, in some of those fields of discovery which lie open to such rough observations as can be made without artificial methods, the great explorers of former times have appropriated most of what is valuable, and that the gleanings which remain are sought after, rather for their abstruseness, than for their intrinsic worth. But the history of science shews that even during that phase of her progress in which she devotes herself to improving the accuracy of the numerical measurement of quantities with which she has long been familiar, she is preparing the materials for the subjugation of new regions, which would have remained unknown if she had been contented with the rough methods of her early pioneers. I might bring forward instances gathered from every branch of science, shewing how the labour of careful measurement has been rewarded by the discovery of new fields of research, and by the development of new scientific ideas. But the history of the science of terrestrial magnetism affords us a sufficient example of what may be done by Experiments in Concert, such as we hope some day to perform in our Laboratory. . . .

But admitting that a practical acquaintance with the methods of Physical Science is an essential part of a mathematical and scientific education, we may be asked whether we are not attributing too much importance to science altogether as part of a liberal education.

Fortunately, there is no question here whether the University should continue to be a place of liberal education, or should devote itself to preparing young men for particular professions. Hence though some of us may, I hope, see reason to make the pursuit of science the main business of our lives, it must be one of our most constant aims to maintain a living connexion between our work and the other liberal studies of Cambridge, whether literary, philological, historical or philosophical.

There is a narrow professional spirit which may grow up among men of science, just as it does among men who practise any other special business. But surely a University is the very place where we should be able to overcome this tendency of men to become, as it were, granulated into small worlds, which are all the more worldly for their very smallness. We lose the advantage of having men of varied pursuits collected into one body, if we do not endeavour to imbibe some of the spirit even of those whose special branch of learning is different from our own.

It is not so long ago since any man who devoted himself to geometry, or to any science requiring continued application, was looked upon as necessarily a misanthrope, who must have abandoned all human interests, and betaken himself to abstractions so far removed from the world of life and action that he has become insensible alike to the attractions of pleasure and to the claims of duty.

In the present day, men of science are not looked upon with the same awe or with the same suspicion. They are supposed to be in league with the material spirit of the age, and to form a kind of advanced Radical party among men of learning.

We are not here to defend literary and historical studies. We admit that the proper study of mankind is man. But is the student of science to be withdrawn from the study of man, or cut off from every noble feeling, so long as he lives in intellectual fellowship with men who have devoted their lives to the discovery of truth, and the results of whose enquiries have impressed themselves on the ordinary speech and way of thinking of men who never heard their names? Or is the student of history and of man to omit from his consideration the history of the origin and diffusion of those ideas which have produced so great a difference between one age of the world and another?

It is true that the history of science is very different from the science of history. We are not studying or attempting to study the working of those blind forces which, we are told, are operating on crowds of obscure people, shaking principalities and powers, and compelling reasonable men to bring events to pass in an order laid down by philosophers.

The men whose names are found in the history of science are not mere hypothetical constituents of a crowd, to be reasoned upon only in masses.

We recognise them as men like ourselves, and their actions and thoughts, being more free from the influence of passion, and recorded more accurately than those of other men, are all the better materials for the study of the calmer parts of human nature.

But the history of science is not restricted to the enumeration of successful investigations. It has to tell of unsuccessful inquiries, and to explain why some of the ablest men have failed to find the key of knowledge, and how the reputation of others has only given a firmer footing to the errors into which they fell. [1871]

Peter Medawar

'The future of man', BBC Reith Lecture, 1959

In this last lecture, I shall discuss the origin in human beings of a new, a non-genetical, system of heredity and evolution based upon certain properties and activities of the brain. The existence of this non-genetical system of heredity is something you are perfectly well aware of. It was not biologists who first revealed to an incredulous world that human beings have brains; that having brains makes a lot of difference; and that a man may influence posterity by other than genetic means. Yet much of what I have read in the writings of biologists seems to say no more than this. I feel a biologist should contribute something towards our *understanding* of the distant origins of human tradition and behaviour, and this is what I shall now attempt. The attempt must be based upon hard thinking, as opposed to soft thinking; I mean, it must be thinking that covers ground and is based upon particulars, as opposed to that which finds its outlet in the mopings or exaltations of poetistic prose.

It will make my argument clearer if I build it upon an analogy. I should like you to consider an important difference between a juke-box and a gramophone – or, if you like, between a barrel-organ and a tape-recorder. A juke-box is an instrument which contains one or more gramophone records, one of which will play whatever is recorded upon it if a particular button is pressed. The act of pressing the button I shall describe as the 'stimulus'. The stimulus is specific: to each button there corresponds one record, and *vice versa*, so that there is a one-to-one relationship between stimulus and response. By pressing a button – any button – I am, in a sense, instructing the juke-box to play music; by pressing this button and not that, I am instructing it to play one piece of music and not another. But – I am not giving the juke-box *musical* instructions. The musical instructions are inscribed upon records that are part of the juke-box, not part of its environment: what a juke-box or barrel-organ can play on any one occasion depends upon structural or inbuilt properties of its own. I shall follow Professor Joshua Lederberg in using the word 'elective' to describe the relationship between what the juke-box plays and the stimulus that impinges upon it from the outside world.

Now contrast this with a gramophone or any other reproducing apparatus. I have a gramophone, and one or more records somewhere in the environment outside it. To hear a particular piece of music, I go through certain motions with switches, and put a gramophone record on. As with the juke-box I am, in a sense, instructing the gramophone to play music, and a particular piece of music. But I am doing more than that: I am giving it musical instructions, inscribed in the grooves of the record I make it play. The gramophone itself contains no source of musical information; it is the record that contains the information, but the record reached the gramophone from the outside world. My relationship to the gramophone – again following Lederberg – I shall describe as 'instructive'; for, in a sense, I *taught* it what to play. With the juke-box, then – and the same goes for a musical-box or barrel-organ – the musical instructions are part of the system that responds to stimuli, and the stimuli are elective: they draw upon the inbuilt capabilities of the instrument. With a gramophone, and still more obviously with a tape recorder, the stimuli are instructive: they endow it with musical capabilities; they import into it musical information from the world outside.

It is we ourselves who have made juke-boxes and gramophones, and who decide what, if anything, they are to play. These facts are irrelevant to the analogy I have in mind, and can be forgotten from now on. Consider only the organism on the one hand – juke-box or gramophone; and, on the other hand, stimuli which impinge upon that organism from the world about it.

During the past ten years, biologists have come to realize that, by and large, organisms are very much more like juke-boxes than gramophones. Most of those reactions of organisms which we were formerly content to regard as instructive are in fact elective. The instructions an organisms contains are not musical instructions inscribed in the grooves of a gramophone record, but *genetical* instructions embodied in chromosomes and nucleic acids. Let me give examples of what I mean.

The oldest example, and the most familiar, concerns the change that comes over a population of organisms when it undergoes an evolution. How should we classify the environmental stimuli that cause organisms to evolve? The Lamarckian theory, the theory that acquired characters can be inherited, is, in its most general form, an *instructive* theory of evolution. It declares that the environment can somehow issue genetical instructions to living organisms – instructions which, duly assimilated, can be passed on from one generation to the next. The blacksmith who is usually called upon to testify on these occasions gets mightily strong arms from forging; somehow this affects the cells that manufacture his spermatozoa, so that his children start life specially well able to develop strong arms. I have no time to explain our tremendous psychological inducement to believe in an instructive or Lamarckian theory of evolution, though in a somewhat more sophisticated form than this. I shall only say that every analysis of what has appeared to be a Lamarckian style of heredity has shown it to be *non*-Lamarckian. So far as we know, the relationship between organism and environment in the evolutionary process is an elective relationship. The environment does *not* imprint genetical instructions upon living things.

Another example: bacteriologists have known for years that if bacteria are forced to live upon some new unfamiliar kind of foodstuff or are exposed to the action of an antibacterial drug, they acquire the ability to make use of that new food, or to make the drug harmless to them by breaking it down. The treatment was at one time referred to as the *training* of bacteria – with the clear implication that the new food or drug *taught* the bacteria how to manufacture the new ferments upon which their new behaviour depends. But it turns out that the process of training belies its name: it is not instructive. A bacterium can synthesize only those ferments which it is genetically entitled to synthesize. The process of training merely brings out or exploits or develops an innate potentiality of the bacterial population, a potentiality underwritten or subsidized by the particular genetic make-up of one or another of its members.

The same argument probably applies to what goes on when animals develop. At one time there was great argument between 'preformationists' and those who believed in epigenesis. The preformationists declared that all development was an unfolding of something already there; the older extremists, whom we now laugh at, believed that a sperm was simply a miniature man. The doctrine of epigenesis, in an equally extreme form, declared that all organisms begin in a homogeneous state, with no apparent or actual structure; and that the embryo is moulded into its adult form solely by stimuli impinging upon it from outside. The truth lies somewhere between these two extreme conceptions. The genetic instructions are preformed, in the sense that they are already there, but their fulfilment is epigenetic – an interpretation that comes close to an elective theory of embryonic development. The environment brings out potentialities present in the embryo in a way which (as with the buttons on a juke-box) is exact and discriminating and specific; but it does not *instruct* the developing embryo in the manufacture of its particular ferments or proteins or whatever else it is made of. Those instructions are already embodied in the embryo: the environment causes them to be carried out.

Until a year or two ago we all felt sure that *one* kind of behaviour indulged in by higher organisms did indeed depend upon the environment as a teacher or instructor. The entry or injection of a foreign substance into the tissues of an animal brings about an immunological reaction. The organism manufactures a specific protein, an 'antibody', which reacts upon the foreign substance, often in such a way as to prevent its doing harm. The formation of antibodies has a great deal to do with resistance to infectious disease. The relationship between a foreign substance and the particular antibody it evokes is exquisitely discriminating and specific; one human being can manufacture hundreds – conceivably thousands – of distinguishable antibodies, even against substances which have only recently been invented, like some of the synthetic chemicals used in industry or in the home. Is the reaction instructive or elective? – *surely*, we all felt, instructive. The organism learns from the chemical pattern of the invading substance just how a particular antibody should be assembled in an appropriate and distinctive way. Self-evident though this interpretation seems, many students of the matter are beginning to doubt it. They hold that the process of forming antibodies is probably elective in character. The information which directs the synthesis of particular antibodies is part of the inbuilt genetical information of the cells that make them; the intruding foreign substance exploits that information and brings it out. It is the juke-box over again. I believe this theory is somewhere near the right one, though I do not accept some of the special constructions that have been put upon it.

So in spite of all will to believe otherwise, and for all that it seems to go against common sense, the picture we are forming of the organism is a juke-box picture – a juke-box containing genetical instructions inscribed upon chromosomes and nucleic acids in much the same kind of way as musical instructions are inscribed upon gramophone records. But what a triumph it would be if an organism could accept information from the environment – if the environment could be made to act in an instructive, not merely an elective, way! A few hundred million years ago a knowing visitor from another universe might have said: 'It's a splendid idea, and I see the point of it perfectly: it would solve – or could solve – the problems of adaptation, and make it possible for organisms to evolve in a much more efficient way than by natural selection. But it's far too difficult: it simply can't be done.'

But you know that it has been done, and that there is just one organ which can accept instruction from the environment: the brain. We know very little about it, but that in itself is evidence of how immensely complicated it is. The evolution of a brain was a feat of fantastic difficulty – the most spectacular enterprise since the origin of life itself. Yet the brain began, I suppose, as a device for responding to elective stimuli. *Instinctive* behaviour is behaviour in which the environment acts electively. If male sex hormones are deliberately injected into a hen, the hen will start behaving in male-like ways. The potentiality for behaving in a male-like manner must therefore have been present in the female; and by pressing (or, as students of behaviour usually say, 'releasing') the right button the environment can bring it out. But the higher parts of the brain respond to instructive stimuli: we *learn*.

Now let me carry the argument forward. It was a splendid idea to evolve into the possession of an organ that can respond to instructive stimuli, but the idea does not go far enough. If that were the whole story, we human beings might indeed live more successfully than other animals; but when we died, a new generation would have to start again from scratch. Let us go back for a moment to genetical instructions. A child at conception receives certain genetical instructions from its parents about how its growth and development are to proceed. Among these instructions there must be some which provide for the issue of further instructions; I mean, a child grows up in such a way that it, too, can eventually have children, and convey genetical instructions to them in turn. We are dealing here with a very special system of communication: a *hereditary* system. There are many examples of systems of this kind. A chain letter is perhaps the simplest: we receive a letter from a correspondent who asks us to write to a third party, asking him in turn to write a letter of the same kind to a fourth, and so on – a hereditary system. The most complicated example is provided by the human brain itself; for it does indeed act as intermediary in a hereditary system of its own. We do more than learn: we teach and hand on; tradition accumulates; we record information and wisdom in books.

Just as a hereditary system is a special kind of system of communication – one in which the instructions provide for the issue of further instructions – so there is a specially important kind of hereditary system: one in which the instructions passed on from one individual to another change in some systematic way in the course of time. A hereditary system with this property may be said to be conducting or undergoing an *evolution*. Genetic systems of heredity often transact evolutionary changes; so also does the hereditary system that is mediated through the brain. I think it is most important to distinguish between four stages in the evolution of a brain. The nervous system began, perhaps, as an organ which responded only to elective stimuli from the environment; the animal that possessed it reacted instinctively or by rote, if at all. There then arose a brain which could begin to accept instructive stimuli from the outside world; the brain in this sense has dim and hesitant beginnings going far back in geological time. The third stage, entirely distinguishable, was the evolution of a non-genetical system of heredity, founded upon the fact that the most complicated brains can do more than merely receive instructions; in one way or another they make it possible for the instructions to be handed on. The existence of this system of heredity – of tradition, in its most general sense – is a defining characteristic of human beings, and it has been important for, perhaps, 500,000 years. In the fourth stage, not clearly distinguishable from the third, there came about a systematic change in the nature of the instructions passed on from generation to generation – an evolution, therefore, and one which has been going at a great pace in the past 200 years. I shall borrow two words used for a slightly different purpose by the great demographer Alfred Lotka to distinguish between the two systems of heredity enjoyed by man: *endosomatic* or internal heredity for the ordinary or genetical heredity we have in common with other animals; and *exosomatic* or external heredity for the non-genetic heredity that is peculiarly our own – the heredity that is mediated through tradition, by which I mean the transfer of information through non-genetic channels from one generation to the next.

I am, of course, saying something utterly obvious: society changes; we pass on knowledge and skills and understanding from one person to another and from one generation to the next; a man can indeed influence posterity by other than genetic means. But I wanted to put the matter in a way which shows that we must not distinguish a strictly biological evolution from a social, cultural, or technological evolution: *both* are biological evolutions: the distinction between them is that the one is genetical and the other is not.

What, then, is to be inferred from all this? What lessons are to be learned from the similarities and correspondences between the two systems of biological heredity possessed by human beings? The answer is important, and I shall now try to justify it: the answer, I believe, is almost none.

It is true that a number of amusing (but in one respect highly dangerous) parallels can be drawn between our two forms of heredity and evolution. Just as biologists speak in a kind of shorthand about the 'evolution' of hearts or ears or legs – it is too clumsy and long-winded to say every time that these organs participate in evolution, or are outward expressions of the course of evolution – so we can speak of the evolution of bicycles or wireless sets or aircraft with the same qualification in mind: they do not really evolve, but they are appendages, exosomatic organs if you like, that evolve with us. And there are many correspondences between the two kinds of evolution. Both are gradual if we take the long view; but on closer inspection we shall find that novelties arise, not everywhere simultaneously – pneumatic tyres did not suddenly appear in the whole population of bicycles – but in a few members of the population: and if these novelties confer economic fitness, or fitness in some more ordinary and obvious sense, then the objects that possess them will spread through the population as a whole and become the prevailing types. In both styles of evolution we can witness an adaptive radiation, a deployment into different environments: there are wireless sets not only for the home, but for use in motor-cars or for carrying about. Some great dynasties die out – airships, for example, in common with the dinosaurs they were so often likened to; others become fixed and stable: toothbrushes retained the same design and constitution for more than a hundred years. And, no matter what the cause of it, we can see in our exosomatic appendages something equivalent to vestigial organs; how else should we describe those functionless buttons on the cuffs of men's coats?

All this sounds harmless enough: why should I have called it dangerous? The danger is that by calling attention to the similarities, which are not profound, we may forget the *differences* between our two styles of heredity and evolution; and the differences between them are indeed profound. In their hunger for synthesis and systematization, the evolutionary philosophers of the nineteenth century and some of their modern counterparts have missed the point: they thought that great lessons were to be learnt from similarities between Darwinian and social evolution; but it is from the differences that all the great lessons are to be learnt. For one thing, our newer style of evolution is Lamarckian in nature. The environment cannot imprint genetical information upon us, but it can and does imprint non-genetical information which we can and do pass on. Acquired characters are indeed inherited. The blacksmith was under an illusion if he supposed that his habits of life could impress themselves upon the genetic make-up of his children; but there is no doubting his ability to teach his children his trade, so that they can grow up to be as stalwart and skilful as himself. It is because this newer evolution is so obviously Lamarckian in character that we are under psychological pressure to believe that genetical evolution must be so too. But although one or two biologists are still feebly trying to graft a Lamarckian or instructive interpretation upon ordinary genetical evolution, they are not nearly so foolish or dangerous as those who have attempted to graft a Darwinian or purely elective interpretation upon the newer, non-genetical, evolution of mankind.

The conception I have just outlined is, I think, a liberating conception. It means that we can jettison all reasoning based upon the idea that changes in society happen in the style and under the pressures of ordinary genetic evolution; abandon any idea that the direction of social change is governed by laws other than laws which have at some time been the subject of human decisions or acts of mind. That competition between one man and another is a necessary part of the texture of society; that societies are organisms which grow and must inevitably die; that division of labour within a society is akin to what we can see in colonies of insects; that the laws of genetics have an overriding authority; that social evolution has a direction forcibly imposed upon it by agencies beyond man's control – all these are biological judgments; but, I do assure you, bad judgments based upon a bad biology. In these lectures you will have noticed that I advocate a 'humane' solution of the problems of eugenics, particularly of the problems of those who have been handicapped by one or another manifestation of the ineptitude of nature. I have not claimed, and do not now claim, that humaneness is an attitude of mind enforced or authorized by some deep inner law of exosomatic heredity: there are technical reasons for supposing that no such laws can exist. I am not warning you against quack biology in order to set myself up as a rival pedlar of patent medicines. What I do say is that our policies and intentions are not to be based upon the supposition that nature knows best; that we are at the mercy of natural laws, and flout them at our peril.

It is a profound truth – realized in the nineteenth century by only a handful of astute biologists and by philosophers hardly at all (indeed, most of those who held any views on the matter held a contrary opinion) – a profound truth that nature does *not* know best; that genetical evolution, if we choose to look at it liverishly instead of with fatuous good humour, is a story of waste, makeshift, compromise, and blunder.

I could give a dozen illustrations of this judgment, but shall content myself with one. You will remember my referring to the immunological defences of the body, the reactions that are set in train by the invasion of the tissues by foreign substances. Reactions of this kind are more than important: they are essential. We can be sure of this because some unfortunate children almost completely lack the biochemical aptitude for making antibodies, the defensive substances upon which so much of resistance to infectious disease depends. Until a few years ago these children died, because only antibiotics like penicillin can keep them alive; for that reason, and because the chemical methods of identifying it have only recently been discovered, the disease I am referring to was only recognized in 1952. The existence of this disease confirms us in our belief that the immunological defences are vitally important; but this does not mean that they are wonders of adaptation, as they are so often supposed to be. Our immunological defences are also an important source of injury, even of mortal injury.

For example: vertebrate animals evolved into the possession of immunological defences long before the coming of mammals. Mammals are viviparous: the young are nourished for some time within the body of the mother: and this (in some ways) admirable device raised for the first time in evolution the possibility that a mother might react immunologically upon her unborn children – might treat them as foreign bodies or as foreign grafts. The haemolytic disease that occurs in about one new-born child in 150 is an error of judgment of just this kind: it is, in effect, an immunological repudiation by the mother of her unborn child. Thus the existence of immunological reactions has not been fully reconciled with viviparity; and this is a blunder – the kind of blunder which, in human affairs, calls forth a question in the House, or even a strongly worded letter to *The Times*.

But this is only a fraction of the tale of woe. Anaphylactic shock, allergy, and hypersensitivity are all aberrations or miscarriages of the immunological process. Some infectious diseases are dangerous to us not because the body fails to defend itself against them but – paradoxically – because it does defend itself: in a sense, the remedy *is* the disease. And within the past few years a new class of diseases has been identified, diseases which have it in common that the body can sometimes react upon its own constituents as if they were foreign to itself. Some diseases of the thyroid gland and some inflammatory diseases of nervous tissue belong to this category; rheumatoid arthritis, lupus erythematosus, and scleroderma may conceivably do so too. I say nothing about the accidents that used to occur in blood transfusions, immunological accidents; nor about the barriers, immunological barriers, that prevent our grafting skin from one person to another, useful though it would so often be; for transfusion and grafting are artificial processes, and, as I said in an earlier lecture, natural evolution cannot be reproached for failing to foresee what human beings might get up to. All I am concerned to show is that natural devices and dispositions are highly fallible. The immunological defences are dedicated to the proposition that anything foreign must be harmful; and this formula is ground out in a totally undiscriminating fashion with results that are sometimes irritating, sometimes harmful, and sometimes mortally harmful. It is far better to have immunological defences than not to have them; but this does not mean that we are to marvel at them as evidences of a high and wise design.

We can, then, improve upon nature; but the possibility of our doing so depends, very obviously, upon our continuing to explore into nature and to enlarge our knowledge and understanding of what is going on. If I were to argue the scientists' case, the case that exploration is a wise and sensible thing to do, I should try to convince you of it by particular reasoning and particular examples, each one of which could be discussed and weighed up; some, perhaps, to be found faulty. I should not say: Man is driven onwards by an exploratory instinct, and can only fulfil himself and his destiny by the ceaseless quest for Truth. As a matter of fact, animals do have what might be loosely called an inquisitiveness, an exploratory instinct; but even if it were highly developed and extremely powerful, it would still not be binding upon us. We should not be *driven* to explore.

Contrariwise, if someone were to plead the virtues of an intellectually pastoral existence, not merely quiet but acquiescent, and with no more than a pensive regret for not understanding what could have been understood; then I believe I could listen to his arguments and, if they were good ones, might even be convinced. But if he were to say that this course of action or inaction was the life that was authorized by Nature; that this was the life Nature provided for and intended us to lead; then I should tell him that he had no proper conception of Nature. People who brandish naturalistic principles at us are usually up to mischief. Think only of what we have suffered from a belief in the existence and overriding authority of a fighting instinct; from the doctrines of racial superiority and the metaphysics of blood and soil; from the belief that warfare between men or classes of men or nations represents a fulfilment of historical laws. These are all excuses of one kind or another, and pretty thin excuses. The inference we can draw from an analytical study of the differences between ourselves and other animals is surely this: that the bells which toll for mankind are – most of them, anyway – like the bells on Alpine cattle; they are attached to our own necks, and it must be *our* fault if they do not make a cheerful and harmonious sound.

'Is the scientific paper a fraud?', from a BBC radio broadcast talk published in 1964

I have chosen for my title a question: Is the scientific paper a fraud? I ought to explain that a scientific 'paper' is a printed communication to a learned journal, and scientists make their work known almost wholly through papers and not through books, so papers are very important in scientific communication. As to what I mean by asking 'is the scientific paper a fraud?' – I do not of course mean 'does the scientific paper misrepresent facts', and I do not mean that the interpretations you find in a scientific paper are wrong or deliberately mistaken. I mean the scientific paper may be a fraud because it misrepresents the processes of thought that accompanied or gave rise to the work that is described in the paper. That is the question, and I will say right away that my answer to it is 'yes'. The scientific paper in its orthodox form does embody a totally mistaken conception, even a travesty, of the nature of scientific thought.

Just consider for a moment the traditional form of a scientific paper (incidentally, it is a form which editors themselves often insist upon). The structure of a scientific paper in the biological sciences is something like this. First, there is a section called the 'introduction' in which you merely describe the general field in which your scientific talents are going to be exercised, followed by a section called 'previous work' in which you concede, more or less graciously, that others have dimly groped towards the fundamental truths that you are now about to expound. Then a section on 'methods' – that is O.K. Then comes the section called 'results'. The section called 'results' consists of a stream of factual information in which it is considered extremely bad form to discuss the significance of the results you are getting. You have to pretend that your mind is, so to speak, a virgin receptacle, an empty vessel, for information which floods into it from the external world for no reason which you yourself have revealed. You reserve all appraisal of the scientific evidence until the 'discussion' section, and in the discussion you adopt the ludicrous pretence of asking yourself if the information you have collected actually means anything; of asking yourself if any general truths are going to emerge from the contemplation of all the evidence you brandished in the section called 'results'.

Of course, what I am saying is rather an exaggeration, but there is more than a mere element of truth in it. The conception underlying this style of scientific writing is that scientific discovery is an inductive process. What induction implies in its cruder form is roughly speaking this: scientific discovery, or the formulation of scientific theory, starts with the unvarnished and unembroidered evidence of the senses. It starts with simple observation – simple, unbiased, unprejudiced, naïve, or innocent observation – and out of this sensory evidence, embodied in the form of simple propositions or declarations of fact, generalizations will grow up and take shape, almost as if some process of crystallization or condensation were taking place. Out of a disorderly array of facts, an orderly theory, an orderly general statement, will somehow emerge. This conception of scientific discovery in which the initiative comes from the unembroidered evidence of the senses was mainly the work of a great and wise, but in this context, I think, very mistaken man – John Stuart Mill.

John Stuart Mill saw, as of course a great many others had seen before him, including Bacon, that deduction in itself is quite powerless as a method of scientific discovery – and for this simple reason: that the process of deduction as such only uncovers, brings out into the open, makes explicit, information that is already present in the axioms or premises from which the process of deduction started. The process of deduction reveals nothing to us except what the infirmity of our own minds has so far concealed from us. It was Mill's belief that induction was the method of science – 'that great mental operation', he called it, 'the operation of discovering and proving general dispositions'. And round this conception there grew up an inductive logic, of which the business was 'to provide rules to which, if inductive arguments conform, those arguments are conclusive'. Now, John Stuart Mill's deeper motive in working out what he conceived to be the essential method of science was to apply that method to the solution of sociological problems: he wanted to apply to sociology the methods which the practice of science had shown to be immensely powerful and exact.

It is ironical that the application to sociology of the inductive method, more or less in the form in which Mill himself conceived it, should have been an almost entirely fruitless one. The simplest application of the Millsian process of induction to sociology came in a rather strange movement called Mass Observation. The belief underlying Mass Observation was apparently this: that if one could only record and set down the actual raw facts about what people do and what people say in pubs, in trains, when they make love to each other, when they are playing games, and so on, then somehow, from this wealth of information, a great generalization would inevitably emerge. Well, in point of fact, nothing important emerged from this approach, unless somebody's been holding out on me. I believe the pioneers of Mass Observation were ornithologists. Certainly they were man-watching – were applying to sociology the very methods which had done so much to bring ornithology into disrepute.

The theory underlying the inductive method cannot be sustained. Let me give three good reasons why not. In the first place, the starting point of induction, naïve observation, innocent observation, is a mere philosophic fiction. There is no such thing as unprejudiced observation. Every act of observation we make is biased. What we see or otherwise sense is a function of what we have seen or sensed in the past.

The second point is this. Scientific discovery or the formulation of the scientific idea on the one hand, and demonstration or proof on the other hand, are two entirely different notions, and Mill confused them. Mill said that induction was the 'operation of discovering and proving general propositions', as if one act of mind would do for both. Now discovery and proof could depend on the same act of mind, and in deduction they do. When we indulge in the process of deduction – as in deducing a theorem from Euclidian axioms or postulates – the theorem contains the discovery (or, more exactly, the uncovery of something which was there in the axioms and postulates, though it was not actually evident) and the process of deduction itself, if it has been carried out correctly, is also the proof that the 'discovery' is valid, is logically correct. So in the process of deduction, discovery and proof can depend on the same process. But in scientific activity they are not the same thing – they are, in fact, totally separate acts of mind.

But the most fundamental objection is this. It simply is not logically possible to arrive with certainty at any generalization containing more information than the sum of the particular statements upon which that generalization was founded, out of which it was woven. How could a mere act of mind lead to the discovery of new information? It would violate a law as fundamental as the law of conservation of matter: it would violate the law of conservation of information.

In view of all these objections, it is hardly surprising that Bertrand Russell in a famous footnote that occurs in his *Principles of Mathematics* of 1903 should have said that, so far as he could see, induction was a mere method of making plausible guesses. And our greatest modern authority on the nature of scientific method, Professor Karl Popper, has no use for induction at all: he regards the inductive process of thought as a myth. 'There is no need even to mention induction,' he says in his great treatise, on *The Logic of Scientific Discovery* – though of course he does.

Now let me go back to the scientific papers. What is wrong with the traditional form of scientific paper is simply this: that all scientific work of an experimental or exploratory character starts with some expectation about the outcome of the inquiry. This expectation one starts with, this hypothesis one formulates, provides the initiative and incentive for the inquiry and governs its actual form. It is in the light of this expectation that some observations are held relevant and others not; that some methods are chosen, others discarded; that some experiments are done rather than others. It is only in the light of this prior expectation that the activities the scientist reports in his scientific papers really have any meaning at all.

Hypotheses arise by guesswork. That is to put it in its crudest form. I should say rather that they arise by inspiration; but in any event they arise by processes that form part of the subject-matter of psychology and certainly not of logic, for there is no logically rigorous method for devising hypotheses. It is a vulgar error, often committed, to speak of 'deducing' hypotheses. Indeed one does not deduce hypotheses: hypotheses are what one deduces things from. So the actual formulation of a hypothesis is – let us say a guess; is inspirational in character. But hypotheses can be tested rigorously – they are tested by experiment, using the word 'experiment' in a rather general sense to mean an act performed to test a hypothesis, that is, to test the deductive consequences of a hypothesis. If one formulates a hypothesis, one can deduce from it certain consequences which are predictions or declarations about what will, or will not, be the case. If these predictions and declarations are mistaken, then the hypothesis must be discarded, or at least modified. If, on the other hand, the predictions turn out correct, then the hypothesis has stood up to trial, and remains on probation as before. This formulation illustrates very well, I think, the distinction between on the one hand the discovery or formulation of a scientific idea or generalization, which is to a greater or lesser degree an imaginative or inspirational act, and on the other hand the proof, or rather the testing of a hypothesis, which is indeed a strictly logical and rigorous process, based upon deductive arguments.

This alternative interpretation of the nature of the scientific process, of the nature of scientific method, is sometimes called the hypothetico-deductive interpretation and this is the view which Professor Karl Popper in the *Logic of Scientific Discovery* has persuaded us is the correct one. To give credit where credit is surely due, it is proper to say that the first professional scientist to express a fully reasoned opinion upon the way scientists actually think when they come upon their scientific discoveries – namely William Whewell, a geologist, and incidentally the Master of Trinity College, Cambridge   was also the first person to formulate this hypothetico-deductive interpretation of scientific activity. Whewell, like his contem-porary Mill, wrote at great length – unnecessarily great length, one is nowadays inclined to think – and I cannot recapitulate his argument, but one or two quotations will make the gist of his thought clear. He said: 'An art of discovery is not possible. We can give no rules for the pursuit of truth which should be universally and peremptorily applicable.' And of hypotheses, he said, with great daring – why it was daring I will explain in just a second – 'a facility in devising hypotheses, so far from being a fault in the intellectual character of a discoverer, is a faculty indispensable to his task'. I said this was daring because the word 'hypothesis' and the conception it stood for was still in Whewell's day a rather discreditable one. Hypotheses had a flavour about them of what was wanton and irresponsible. The great Newton, you remember, had frowned upon hypotheses. '*Hypotheses non fingo*', he said, and there is another version in which he says '*hypotheses non sequor*' – I do not pursue hypotheses.

So to go back once again to the scientific paper: the scientific paper is a fraud in the sense that it does give a totally misleading narrative of the processes of thought that go into the making of scientific discoveries. The inductive format of the scientific paper should be discarded. The discussion which in the traditional scientific paper goes last should surely come at the beginning. The scientific facts and scientific acts should follow the discussion, and scientists should not be ashamed to admit, as many of them apparently *are* ashamed to admit, that hypotheses appear in their minds along uncharted by-ways of thought; that they are imaginative and inspirational in character; that they are indeed adventures of the mind. What, after all, is the good of scientists reproaching others for their neglect of, or indifference to, the scientific style of thinking they set such great store by, if their own writings show that they themselves have no clear understanding of it?

Anyhow, I am practising what I preach. What I have said about the nature of scientific discovery you can regard as being itself a hypothesis, and the hypothesis comes where I think it should be, namely, it comes at the beginning of the series. Later speakers will provide the facts which will enable you to test and appraise this hypothesis, and I think you will find – I hope you will find – that the evidence they will produce about the nature of scientific discovery will bear me out.

'On "The effecting of all things possible" ', from *Pluto's Republic*, 1982

My title, or, if you like, my motto, comes from Francis Bacon's *New Atlantis*, published in 1627. The *New Atlantis* was Bacon's dream of what the world might have been, and might still become, if human knowledge were directed towards improving the worldly condition of man. It makes a rather strange impression nowadays, and very few people bother with it who are not interested either in Bacon himself, or in the flux of seventeenth-century opinion or the ideology of Utopias. We shall not read it for its sociological insights, which are non-existent, nor as science fiction, because it has a general air of implausibility; but there is one high poetic fancy in the *New Atlantis* that stays in the mind after all its fancies and inventions have been forgotten. In the New Atlantis, an island kingdom lying in very distant seas, the only commodity of external trade is – *light*: Bacon's own special light, the light of understanding. The Merchants of Light who carry out its business are members of a society or order of philosophers who between them make up (so their spokesman declares) 'the noblest foundation that ever was upon the earth'. 'The end of our foundation', the spokesman goes on to say, 'is the knowledge of causes and the secret motions of things; and the enlarging of the bounds of human empire, to the effecting of all things possible.' You will see later on why I chose this motto.

My purpose is to draw certain parallels between the spiritual or philosophic condition of thoughtful people in the seventeenth century and in the contemporary world, and to ask why the great philosophic revival that brought comfort and a new kind of understanding to our predecessors has now apparently lost its power to reassure us and cheer us up.

The period of English history that lies roughly between the accession of James I in 1603 and the English Civil War has much in common with the present day. . . . For the historian of ideas, it is a period of questioning and irresolution and despondency; of sermonising but also of satire; of rival religions competing for allegiance, among them the 'black doctrine of absolute reprobation'; a period during which our human propensity towards hopefulness was clouded over by a sense of inconstancy and decay. Literary historians have spoken of a 'metaphysical shudder', . . . and others of a sense of crisis or of a 'failure of nerve'. . . . Of course, we must not imagine that ordinary people went around with the long sunk-in faces to be expected in the victims of a spiritual deficiency disease. It was philosophic or reflective man who had these misgivings, the man who is all of us some of the time but none of us all of the time, and we may take it that, then as now, the remedy for discomforting thoughts was less often to seek comfort than to abstain from thinking.

Amidst the philosophic gloom of the period I am concerned with, new voices began to be heard which spoke of hope and of the possibility of a future (a subject I shall refer to later on); which spoke of confidence in human reason, and of what human beings might achieve through an understanding of nature and a mastery of the physical world. I think there can be no question that, in this country, it was Francis Bacon who started the dawn chorus – the man who first defined the newer purposes of learning and, less successfully, the means by which they might be fulfilled. Human spirits began to rise. To use a good old seventeenth-century metaphor, there was a slow change, but ultimately a complete one, in the 'climate of opinion'. It became no longer the thing to mope. In a curious way the Pillars of Hercules – the 'Fatal Columns' guarding the Straits of Gibraltar that make the frontispiece to Bacon's *Great Instauration* – provided the rallying cry of the New Philosophy. Let me quote a great American scholar's, Dr Marjorie Hope Nicolson's, . . . description of how this came about:

> Before Columbus set sail across the Atlantic, the coat of arms of the Royal Family of Spain had been an *impressa*, depicting the Pillars of Hercules, the Straits of Gibraltar, with the motto, *Ne Plus Ultra*. There was 'no more beyond'. It was the glory of Spain that it was the outpost of the world. When Columbus made his discovery, Spanish Royalty thriftily did the only thing necessary: erased the negative, leaving the Pillars of Hercules now bearing the motto, *Plus Ultra*. There was more beyond . . .

And so *plus ultra* became the motto of the New Baconians, and the frontispiece to the *Great Instauration* shows the Pillars of Hercules with ships passing freely to and fro.

One symptom of the new spirit of enquiry was, of course, the foundation of the Royal Society and of sister academies in Italy and France. That story has often been told, and in more than one version, because the parentage of the Royal Society is still in question. . . . We shall be taking altogether too narrow a view of things, however, if we suppose that the great philosophic uncertainties of the seventeenth century were cleared up by the fulfilment of Bacon's ambitions for science. Modern scientific research began earlier than the seventeenth century. . . . The great achievement of the latter half of the seventeenth century was to arrive at a general scheme of belief within which the cultivation of science was seen to be very proper, very useful, and by no means irreligious. This larger conception or purpose, of which science was a principal agency, may be called 'rational humanism' if we are temperamentally in its favour and take our lead from the writings of John Locke, or 'materialistic rationalism' if we are against it and frown disapprovingly over Thomas Hobbes, but neither description is satisfactory, because the new movement had not yet taken on the explicit character of an alternative or even an antidote to religion, which is the sense that 'rational humanism' tends to carry with it today.

However we may describe it, rational humanism became the dominant philosophic influence in human affairs for the next 150 years, and by the end of the eighteenth century the spokesmen of Reason and Enlightenment – men such as Adam Ferguson and William Godwin and Condorcet – take completely for granted many of the ideas that had seemed exhilarating and revolutionary in the century before. But over this period an important transformation was taking place. The seventeenth-century doctrine of the *necessity* of reason was slowly giving way to a belief in the *sufficiency* of reason – so illustrating the tendency of many powerful human beliefs to develop into an extreme or radical form before they lose their power to persuade us, and in doing so to create anew many of the evils for which at one time they professed to be the remedy. (It has often been said that rationalism in its more extreme manifestations could only supplant religion by acquiring some of the characteristics of religious belief itself.) Please don't interpret these remarks as any kind of attempt to depreciate the power of reason. I emphasise the distinction between the ideas of the necessity and of the sufficiency of reason as a defence against that mad and self-destructive form of anti-rationalism which seems to declare that because reason is not sufficient, it is not necessary.

Many reflective people nowadays believe that we are back in the kind of intellectual and spiritual turmoil that disturbed the first half of the seventeenth century. Both epochs are marked, not by any characteristic system of beliefs (neither can be called 'The Age of' anything), but by an equally characteristic syndrome of unfixed beliefs; by the emptiness that is left when older doctrines have been found wanting and none has yet been found to take their place. Both epochs have the characteristics of a philosophic interregnum. In the first half of the seventeenth century, the essentially medieval world-picture of Elizabethan England had lost its power to satisfy and bring comfort, just as nowadays the radical materialism traditionally associated with Victorian thinkers seems quite inadequate to remedy our complaints. By a curious inversion of thinking, scholastic reasoning is said to have failed because it discouraged new enquiry, but that was precisely the measure of its success. For that is just what successful, satisfying explanations do: they confer a sense of finality; they remove the incentive to work things out anew. At all events the repudiation of Aristotle and the hegemony of ancient learning, of the scholastic style of reasoning, of the illusion of a Golden Age, is as commonplace in the writings of the seventeenth century as dismissive references to rationalism and materialism are in the literature of the past fifty years.

We can draw quite a number of detailed correspondences between the contemporary world and the first forty or fifty years of the seventeenth century, all of them part of a syndrome of dissatisfaction and unbelief; and though we might find reason to cavil at each one of them individually, they add up to an impressive case. Novels and philosophical *belles-lettres* have now an inward-looking character, a deep concern with matters of personal salvation and a struggle to establish the authenticity of personal existence; and we may point to the prevalence of satire and of the Jacobean style of 'realism' on the stage. I shall leave aside the political and economic correspondences between the two epochs, . . . important though they are, and confine myself to analogies that might be described as 'philosophical' in the homely older sense, the sense that has to do with the purpose and conduct of life and with the attempt to answer the simple questions that children ask. Once again we are oppressed by a sense of decay and deterioration, but this time, in part at least, by a fear of the deterioration of the world through technological innovation. Artificial fertilisers and pesticides are undermining our health (we tell ourselves), soil and sea are being poisoned by chemical and radioactive wastes, drugs substitute one kind of disease for another, and modern man is under the influence of stimulants whenever he is not under the influence of sedatives. Once again there is a feeling of despondency and incompleteness, a sense of doubt about the adequacy of man, amounting in all to what a future historian might again describe as a failure of nerve. Intelligent and learned men may again seek comfort in an elevated kind of barminess (but something kind and gentle nevertheless). Mystical syntheses between science and religion, like the Cambridge Neoplatonism of the mid-seventeenth century, have their counterpart today, perhaps, in the writings and cult of Teilhard de Chardin and in a revival of faith in the Wisdom of the East. Once again there is a rootlessness or ambivalence about philosophical thinking, as if the discovery or rediscovery of the insufficiency of reason had given a paradoxical validity to nonsense, and this gives us a special sympathy for the dilemmas of the seventeenth century. To William Lecky, the great nineteenth-century historian of rationalism, it seemed almost beyond comprehension that witch hunting and witch burning should have persisted far into the seventeenth century, or that Joseph Glanvill should have been equally an advocate of the Royal Society and of belief in witchcraft. . . .

We do not wonder at it now. It no longer seems strange to us that Pascal the geometer who spoke with perfect composure about infinity and the infinitesimal should have been supplanted by Pascal the great cosmophobe who spoke with anguish about the darkness and loneliness of outer space. Discoveries in astronomy and cosmology have always a specially disturbing quality. We remember the dismay of John Donne and Pascal himself and latterly of William Blake. Cosmological discoveries bring with them a feeling of awe but also, for most people, a sense of human diminishment. Our great sidereal adventures today are both elevating and frightening, and may be both at the same time. The launching of a space rocket is (to go back to seventeenth-century language) a tremendous phenomenon. It must have occurred to many who saw pictures of it that the great steel rampart or nave from which the Apollo rockets were launched had the size and shape and grandeur of a cathedral, with Apollo itself in the position of a spire. Like a cathedral it is economically pointless, a shocking waste of public money; but like a cathedral it is also a symbol of aspiration towards higher things.

When we compare the climates of opinion in the seventeenth century and today, we must again remember that cries of despair are not necessarily authentic. There was a strong element of affectation about Jacobean melancholy, and so there is today. Then as now it had tended to become a posture. One of a modern writer's claims to be taken seriously is to castigate complacency and to show up contentment for the shallow and insipid thing that it is assumed to be. But ordinary human beings continue to be vulgarly high spirited. The character we all love best in Boswell is Johnson's old college companion, Mr Oliver Edwards – the man who said that he had tried in his time to be a philosopher, but had failed because cheerfulness was always breaking in.

I should now like to describe the new style of thinking that led to a great revival of spirits in the seventeenth century. It is closely associated with the birth of science, of course – of Science with a capital S – and the 'new philosophy' that had been spoken of since the beginning of the century referred to the beginnings of physical science; but (as I said a little earlier) we should be taking too narrow a view of things if we supposed that the instauration of science made up the whole or even the greater part of it. The new spirit is to be thought of not as scientific, but as something conducive to science; as a movement within which scientific enquiry played a necessary and proper part.

What then were the philosophic elements of the new revival (using 'philosophy' again in its homely sense)?

The seventeenth century was an age of Utopias, though Thomas More's own Utopia was already years old. The Utopias or anti-Utopias we devise today are usually set in the future, partly because the world's surface is either tenanted or known to be empty, partly because we need and assume we have time for the fulfilment of our designs. The old Utopias – Utopia itself, the New Atlantis, Christianopolis, and the City of the Sun . . . – were contemporary societies. Navigators and explorers came upon them accidentally in far-off seas. What is the meaning of the difference? One reason, of course, is that the world then still had room for undiscovered principalities, and geographical exploration itself had the symbolic significance we now associate with the great adventures of modern science. Indeed, now that outer space is coming to be our playground, we may again dream of finding ready-made Utopias out there. But this is not the most important reason. The old Utopias were not set in the future because very few people believed that there would *be* a future – an earthly future, I mean; nor was it by any means assumed that the playing-out of earthly time would improve us or increase our capabilities. On the contrary, time was running out, in fulfilment of the great Judaic tradition, and we ourselves were running down.

These thoughts suffuse the philosophic speculation of the seventeenth century until quite near its end. 'I was borne in the last Age of the World,' said John Donne, . . . and Thomas Browne speaks of himself as one whose generation was 'ordained in this setting of time'. . . . The most convincing evidence of the seriousness of this belief is to be found not in familiar literary tags, but in the dull and voluminous writings of those who, like George Hakewill, . . . repudiated the idea of human deterioration and the legend of a golden age, but had no doubt at all about the imminence of the world's end. The apocalyptic forecast was, of course, a source of strength and consolation to those who had no high ambitions for life on earth. The precise form the end of history would take had long been controversial – the New Jerusalem might be founded upon the earth itself or be inaugurated in the souls of men in heaven – but that history would come to an end had hardly been in question. Towards the end of the sixteenth century there had been some uneasy discussion of the idea that the material world might be eternal, but the thought had been a disturbing one, and had been satisfactorily explained away. . . .

During the seventeenth century this attitude changes. The idea of an end of history is incompatible with a new feeling about the great things human beings might achieve through their own ingenuity and exertions. The idea therefore drops quietly out of the common consciousness. It is not refuted, but merely fades away. It is true that the idea of human deterioration was expressly refuted – in England by George Hakewill but before him by Jean Bodin (by whom Hakewill was greatly influenced) and by Louis le Roy. . . . The refutation of the idea of decay did not carry with it an acceptance of the idea of progress, or anyhow of linear progress: it was a question of recognising that civilisations or cultures had their ups and downs, and went through a life cycle of degeneration and regeneration – a 'circular kind of progress', Hakewill said.

There were, however, two elements of seventeenth-century thought that imply the idea of progress even if it is not explicitly affirmed. The first was the recognition that the tempo of invention and innovation was speeding up, that the flux of history was becoming denser. In *The City of the Sun* Campanella tells us that 'his age has in it more history within a hundred years than all the world had in four thousand years before it'. He is echoing Peter Ramus: . . . 'We have seen in the space of one age a more plentiful crop of learned men and works than our predecessors saw in the previous fourteen.' By the latter half of the seventeenth century the new concept had sunk in.

The second element in the concept of futurity – in the idea that men might look forward, not only backwards or upwards – is to be found in the breathtaking thought that there was no apparent limit to human inventiveness and ingenuity. It was the notion of a perpetual *plus ultra*, that what was already known was only a tiny fraction of what remained to be discovered, so that there would always be more beyond. Bacon published his *Novum organum* at the beginning of the remarkable decade between 1620 and 1630, and had singled it out as the greatest obstacle to the growth of understanding, that 'men despair and think things impossible'. 'The human understanding is unquiet', he wrote; 'it cannot stop or rest and still presses onwards, but in vain' – in vain, because our spirits are oppressed by 'the obscurity of nature, the shortness of life, the deceitfulness of the senses, the infirmity of judgement, the difficulty of experiment, and the like'. 'I am now therefore to speak of hope', he goes on to say, in a passage that sounds like the trumpet calls in *Fidelio*. The hope he held out was of a rebirth of learning, and with it the realisation that if men would only concentrate and direct their faculties, 'there is no difficulty that might not be overcome'. '[T]he process of Art is indefinite,' wrote Henry Power, 'and who can set a *non-ultra* to her endeavours?' . . . There is a mood of exultation and glory about this new belief in human capability and the future in which it might unfold. With Thomas Hobbes 'glorying' becomes almost a technical term: 'Joy, arising from imagination of a man's own power and ability, is that exultation of mind called glorying', he says, in *Leviathan*, and in another passage he speaks of a perseverance of delight in the continual and indefatigable generation of knowledge'.

It does not take a specially refined sensibility to see how exciting and exhilarating these new notions must have been. During the eighteenth century, of course, everybody sobers up. The idea of progress is taken for granted – but in some sense it gets out of hand, for not only will human inventions improve without limit, but so also (it is argued, though not very clearly) will human beings. It is interesting to compare the exhilaration of the seventeenth century with, say, William Godwin's magisterial tone of voice as the eighteenth century draws to an end. 'The extent of our progress in the cultivation of human knowledge is unlimited. Hence it follows . . . [t]hat human inventions . . . are susceptible of perpetual improvement.'

> Can we arrest the progress of the inquiring mind? If we can, it must be by the most unmitigated despotism. Intellect has a perpetual tendency to proceed. It cannot be held back, but by a power that counteracts its genuine tendency, through every moment of its existence. Tyrannical and sanguinary must be the measures employed for this purpose. Miserable and disgustful must be the scene they produce. . . . [William Godwin, *An Enquiry Concerning Political Justice*]

The seventeenth century had begun with the assumption that a powerful force would be needed to put the inventive faculty into motion; by the end of the eighteenth century it is assumed that only the application of an equally powerful force could possibly slow it down.

Before going on, it is worth asking if this conception is still acceptable – that the growth of knowledge and know-how has no intrinsic limit. We have now grown used to the idea that most ordinary or natural growth processes (the growth of organisms or populations of organisms or, for example, of cities) is not merely limited, but self-limited, that is, is slowed down and eventually brought to a standstill *as a consequence of the act of growth itself*. For one reason or another, but always for some reason, organisms cannot grow indefinitely, just as beyond a certain level of size or density a population defeats its own capacity for further growth. May not the body of knowledge also become unmanageably large, or reach such a degree of complexity that it is beyond the comprehension of the human brain? To both these questions I think the answer is 'No'. The proliferation of recorded knowledge and the seizing-up of communications pose technological problems for which technical solutions can and are being found. As to the idea that knowledge may transcend the power of the human brain: in a sense it has long done so. No one can 'understand' a radio-set or automobile in the sense of having an effective grasp of more than a fraction of the hundred technologies that enter into their manufacture. But we must not forget the additiveness of human capabilities. We work through consortia of intelligences, past as well as present. We might, of course, blow ourselves up or devise an unconditionally lethal virus, but we don't *have* to. Nothing of the kind is necessarily entailed by the growth of knowledge and understanding. I do not believe that there is any intrinsic limitation upon our ability to answer the questions that belong to the domain of natural knowledge and fall therefore within the agenda of scientific enquiry.

The repudiation of the concept of decay, the beginnings of a sense of the future, an affirmation of the dignity and worthiness of secular learning, the idea that human capabilities might have no upper limit, an exultant recognition of the capabilities of man – these were the seventeenth century's antidote to despondency. You may wonder why I have said nothing about the promulgation of the experimental method in science as one of the decisive intellectual movements of the day. My defence is that the origin of the experimental method has been the subject of a traditional misunderstanding, the effect of reading into the older usages of 'experiment' the very professional meaning we attach to that word today. Bacon is best described as an advocate of the *experiential* method in science – of the belief that natural knowledge was to be acquired not from authority, however venerable, nor by syllogistic exercises, however subtle, but by paying attention to the evidence of the senses, evidence from which (he believed) all deception and illusion could be stripped away. Bacon's writings form one of the roots of the English tradition of philosophic empiricism, of which the greatest spokesman was John Locke. The unique contribution of science to empirical thought lay in the idea that experience could be *stretched* in such a way as to make nature yield up information which we should otherwise have been unaware of. The word 'experiment' as it was used until the nineteenth century stood for the concept of stretched or deliberately contrived experience; for the belief that we might make nature perform according to a scenario of our own choosing instead of merely watching her own artless improvisations. An 'experiment' today is not something that merely enlarges our sensory experience. It is a critical operation of some kind that discriminates between hypotheses and therefore gives a specific direction to the flow of thought. Bacon's championship of the idea of experimentation was part of a greater intellectual movement which had a special manifestation in science without being distinctively scientific. His reputation should not, and fortunately need not, rest on his being the founder of the 'experimental method' in the modern sense. . . .

Let us return to the contemporary world and discuss our misgivings about the way things are going now. No one need suppose that our present philosophic situation is unique in its character and gravity. It was partly to dispel such an illusion that I have been moving back and forth between the seventeenth century and the present day. Moods of complacency and discontent have succeeded each other during the past 400 or 500 years of European history, and our present mood of self-questioning does not represent a new and startled awareness that civilisation is coming to an end. On the contrary, the existence of these doubts is probably our best assurance that civilisation will continue.

Many of the ingredients of the seventeenth-century antidote to melancholy have lost their power to bring peace of mind today, and have become a source of anxiety in themselves. Consider the tempo of innovation. In the post-Renaissance world the feeling that inventiveness was increasing and that the whole world was on the move did much to dispel the myth of deterioration and give people confidence in human capability. Nevertheless the tempo was a pretty slow one, and technical innovation had little influence on the character of common life. A man grew up and grew old in what was still essentially the world of his childhood; it had been his father's world and it would be his children's too. Today the world changes so quickly that in growing up we take leave not just of youth but of the world we were young in. I suppose we all realise the degree to which fear and resentment of what is new is really a lament for the memories of our childhood. Dear old steam trains, we say to ourselves, but nasty diesel engines; trusty old telegraph poles but horrid pylons. Telegraph poles, as a Poet Laureate told us a good many years ago, . . . are something of a test case. Anyone who has spent part of his childhood in the countryside can remember looking up through the telegraph wires at a clouded sky and discerning the revolution of the world, or will have listened, ear to post, to the murmur of interminable conversations. For some people even the smell of telegraph poles is nostalgic, though creosote has a pretty technological smell. Telegraph poles have been assimilated into the common consciousness, and one day pylons will be, too. When the pylons are dismantled and the cables finally go underground, people will think again of those majestic catenary curves, and remind each other of how steel giants once marched across the countryside in dead silence and in single file. (What is wrong with pylons is that most of them are ugly. If only the energy spent in denouncing them had been directed towards improving their appearance, they could have been made as beautiful, even as majestic, as towers or bridges are allowed to be, and need not have looked incongruous in the countryside.)

When Bacon described himself as a trumpeter of the new philosophy, the message he proclaimed was of the virtue and dignity of scientific learning and of its power to make the world a better place to live in. I am continually surprised by the superficiality of the reasons which have led people to question those beliefs today. Many different elements enter into the movement to depreciate the services to mankind of science and technology. I have just mentioned one of them, the tempo of innovation when measured against the span of life. We wring our hands over the miscarriages of technology and take its benefactions for granted. We are dismayed by air pollution but not proportionately cheered up by, say, the virtual abolition of poliomyelitis. (Nearly 5,000 cases of poliomyelitis were recorded in England and Wales in 1957. In 1967 there were less than thirty.) There is a tendency, even a perverse willingness, to suppose that the despoliation sometimes produced by technology is an inevitable and irremediable process, a trampling down of nature by the big machine. Of course it is nothing of the kind. The deterioration of the environment produced by technology is a technological problem for which technology has found, is finding, and will continue to find solutions. There is, of course, a sense in which science and technology can be arraigned for devising new instruments of warfare, but another and more important sense in which it is the height of folly to blame the weapon for the crime. I would rather put it this way: in the management of our affairs we have too often been bad workmen, and like all bad workmen we blame our tools. I am all in favour of a vigorously critical attitude towards technological innovation: we should scrutinise all attempts to improve our condition and make sure that they do not in reality do us harm; but there is all the difference in the world between informed and energetic criticism and a drooping despondency that offers no remedy for the abuses it bewails.

Superimposed on all particular causes of complaint is a more general cause of dissatisfaction. Bacon's belief in the cultivation of science for the 'merit and emolument of life' has always been repugnant to those who have taken it for granted that comfort and prosperity imply spiritual impoverishment. But the real trouble nowadays has very little to do with material prosperity or technology or with our misgivings about the power of research and learning generally to make the world a better place. The real trouble is our acute sense of human failure and mismanagement, a new and specially oppressive sense of the inadequacy of men. So much was hoped of us, particularly in the eighteenth century. We were going to improve, weren't we? – and for some reason which was never made clear to us we were going to grow in moral stature as well as in general capability. Our school reports were going to get better term by term. Unfortunately they haven't done so. Every folly, every enormity that we look back on with repugnance can find its equivalent in contemporary life. Once again our intellectuals have failed us; there is a general air of misanthropy and self-contempt, of protest, but not of affirmation. There is a peculiar selfishness about modern philosophic speculation (using 'philosophy' here again in its homely or domestic sense). The philosophic universe has contracted into a neighbourhood, a suburbia of personal relationships. It is as if the classical formula of self-interest, 'I'm all right, Jack', was seeking a new context in our private, inner world.

We can obviously do better than this, and there is just one consideration that might help to take the sting out of our self-reproaches. In the melancholy reflections of the post-Renaissance era it was taken for granted that the poor old world was superannuated, that history had all but run its course and was soon coming to an end. The brave spirits who inaugurated the new science dared to believe that it was *not* too late to be ambitious, but now we must try to understand that it is a bit too early to expect our grander ambitions to be fulfilled. Today we are conscious that human history is only just beginning. There has always been room for improvement; now we know that there is time for improvement, too. For all their intelligence and dexterity – qualities we have always attached great importance to – the higher primates (monkeys, apes and men) have not been very successful. Human beings have a history of more than 500,000 years. Only during the past 5,000 years or thereabouts have human beings won a reward for their special capabilities; only during the past 500 years or so have they begun to be, in the biological sense, a success. If we imagine the evolution of living organisms compressed into one year of cosmic time, then the evolution of man has occupied a day. Only during the past ten to fifteen minutes of the human day has our life on earth been anything but precarious. Until then we might have gone under altogether or, more likely, have survived as a biological curiosity; as a patchwork of local communities only just holding their own in a bewildering and hostile world. Only during this past fifteen minutes (for reasons I shall not go into, though I think they can be technically explained) has there been progress, though, of course, it doesn't amount to very much. We cannot point to a single definitive solution of any one of the problems that confront us – political, economic, social or moral, that is, having to do with the conduct of life. We are still beginners, and for that reason may hope to improve. To deride the hope of progress is the ultimate fatuity, the last word in poverty of spirit and meanness of mind. There is no need to be dismayed by the fact that we cannot yet envisage a definitive solution of our problems, a resting-place beyond which we need not try to go. Because he likened life to a race, . . . and defined felicity as the state of mind of those in the front of it, Thomas Hobbes has always been thought of as the arch materialist, the first man to uphold go-getting as a creed. but that is a travesty of Hobbes's opinion. He was a go-getter in a sense, but it was the going, not the getting, he extolled. As Hobbes conceived it, the race had no finishing post. The great thing about the race was to be in it, to be a contestant in the attempt to make the world a better place, and it was a spiritual death he had in mind when he said that to forsake the course is to die. 'There is no such thing as perpetual tranquillity of mind while we live here', he told us in *Leviathan*, 'because life itself is but motion and can never be without desire, or without fear, no more than without sense'; 'there can be no contentment but in proceeding'. I agree.

Michael Newman

'Cloned poem', published in *The Sciences*, 1982

The Flea

Marke but this flea, and marke in this,
How little that which thou deny'st me is;
It suck'd me first, and now sucks thee,
And in this flea, our two bloods mingled bee;
Thou know'st that this cannot be said
A sinne, nor shame, nor losse of maidenhead,
   Yet this enjoyes before it wooe,
   And pamper'd swells with one blood made of two,
   And this, alas, is more than wee would doe.

Oh stay, three lives in one flea spare,
Where wee almost, yea more than maryed are.
This flea is you and I, and this
Our marriage bed, and marriage temple is;
Though parents grudge, and you, w'are met,
And cloysterd in these living walls of Jet.
   Though use make you apt to kill mee,
   Let not to that, selfe murder added bee,
   And sacrilege, three sinnes in killing three.

Cruell and sodaine, hast thou since
Purpled thy naile, in blood of innocence?
Wherein could this flea guilty bee,

Except in that drop which it suckt from thee?
Yet thou triumph'st, and saist that thou
Find'st not thy selfe, nor mee the weaker now;
  'Tis true, then learne how false, feares bee;
  Just so much honor, when thou yeeld'st to mee,
  Will wast, as this flea's death tooke life from thee.

John Donne

The Virus

Observe this virus: think how small
Its arsenal, and yet how loud its call;
It took my cell, now takes your cell,
And when it leaves will take our genes as well.
Genes that are master keys to growth
That turn it on, or turn it off, or both;
  Should it return to me or you
  It will own the skeleton keys to do
  A number on our tumblers; stage a coup.

But would you kill the us in it,
The sequence that it carries, bit by bit?
The virus was the first to live,
Or lean in that direction; now we give
Attention to its way with locks,
And how its tickings influence our clocks:
  Its gears fit in our clockworking,
  Its habits of expression have a ring
  That makes our carburetors start to ping.

This happens when cells start to choke
As red cells must in monoxidic smoke,
Where membranes get the guest-list wrong
And single-file becomes a teeming throng,
And growth exists for its own sake:
Then soon enough the healthy genes must break;
  If we permit this with our cells,
  With molecules abet the clanging bells:
  Lend our peculiar tone to our death knells.

Michael Newman

Isaac Newton

An extract from a letter to Hawes, 1694, from volume III of the *Correspondence*

And now I have told you my opinion in these things, I will give you Mr. Oughtred's, a Man whose judgment (if any man's) may be safely relyed upon. For he in his book of the circles of proportion, in the end of what he writes about Navigation (page 184) has this exhortation to Seamen 'And if, saith he, the Masters of Ships and Pilots will take the pains in the Journals of their Voyages diligently & faithfully to set down in severall columns, not onely the Rumb they goe on and the measure of the Ships way in degrees, & the observation of Latitude and variation of their compass; but alsoe their conjectures and reason of their correction they make of the aberrations they shall find, and the qualities & condition of their ship, and the diversities and seasons of the winds, and the secret motions or agitations of the Seas, when they begin, and how long they continue, how farr they extend & wth what inequality; and what else they shall observe at Sea worthy consideration, & will be pleased freely to communicate the same with Artists, such as are indeed skilfull in the Mathematicks and lovers & enquirers of the truth: I doubt not but that there shall be in convenient time, brought to light many necessary precepts wch may tend to ye perfecting of Navigation, and the help and safety of such whose Vocations doe inforce them to commit their lives and estates in the vast Ocean to the providence of God.' Thus farr that very good and judicious man Mr. Oughtred. I will add, that if instead of sending the Observations of Seamen to able Mathematicians at Land, the Land would send able Mathematicians to Sea, it would signify much more to the improvement of Navigation and safety of Mens lives and estates on that element.

Norman Nicholson

An extract from *Windscale*, from *A Local Habitation*, 1972

> The toadstool towers infest the shore:
> Stink-horns that propagate and spore
> Wherever the wind blows.
> Scafell looks down from the bracken band,
> And sees hell in a grain of sand,
> And feels the canker itch between his toes.
>   This is a land where dirt is clean,
> And poison pasture, quick and green,
> And storm sky, bright and bare;
> Where sewers flow with milk, and meat
> Is carved up for the fire to eat,
> And children suffocate in God's fresh air.

Michael Polanyi

An extract from *Journeys in Belief*, 1958

Objectivism has totally falsified our conception of truth, by exalting what we can know and prove, while covering up with ambiguous utterances all that we know and *cannot* prove, even though the latter knowledge underlies and must ultimately set its seal to, all that we *can* prove. In trying to restrict our minds to the few things that are demonstrable, and therefore explicitly dubitable, it has overlooked the a-critical choices which determine the whole being of our minds, and has rendered us incapable of acknowledging these vital choices.

John Polkinghorne

'Perplexities', chapter 1 of *The Quantum World*

A layman venturing into the quantum world no doubt expects to encounter some fairly strange phenomena. He is prepared for the paradoxical. Yet the greatest paradox of all is likely to escape his attention unless he has a candid professional friend to point it out to him. It is simply this. Quantum theory is both stupendously successful as an account of the small-scale structure of the world and it is also the subject of unresolved debate and dispute about its interpretation. That sounds rather like being shown an impressively beautiful palace and being told that no one is quite sure whether its foundations rest on bedrock or shifting sand.

Concerning the successfulness of quantum theory there can be no dissent. From the time it reached its fully articulated form in the middle twenties of this century it has been used daily by an army of honest toilers with consistently reliable results. Originally constructed to account for atomic physics it has proved equally applicable to the behaviour of those latest candidates for the role of basic constituents of matter, the quarks and gluons. In going from atoms to quarks there is a change of scale by a factor of at least ten million. It is impressive that quantum mechanics can take that in its stride.

The problems of interpretation cluster around two issues: the nature of reality and the nature of measurement. Philosophers of science have latterly been busy explaining that science is about correlating phenomena or acquiring the power to manipulate them. They stress the theory-laden character of our pictures of the world and the extent to which scientists are said to be influenced in their thinking by the social factor of the spirit of the age. Such accounts cast doubt on whether an understanding of reality is to be conceived of as the primary goal of science or the actual nature of its achievement. These comments from the touchline may well contain points of value about the scientific game. They should not, however, cause us to neglect the observations of those who are actually players. The overwhelming impression of the participants is that they are investigating the way things are. Discovery is the name of the game. The pay-off for the rigours and *longueurs* of scientific research is the consequent gain in understanding of the way the world is constructed. Contemplating the sweep of the development of some field of science can only reinforce that feeling.

Consider, for example, our understanding of electricity and magnetism and the nature of light. In the nineteenth century, first Thomas Young demonstrated the wave character of light; then Faraday's brilliant experimental researches revealed the interlocking nature of electricity and magnetism; finally the theoretical genius of Maxwell produced an understanding of the electromagnetic field whose oscillations were identifiable with Young's light waves. It all constituted a splendid achievement. Nature, however, proved more subtle than even Maxwell had imagined. The beginning of this century produced phenomena which equally emphatically showed that light was made up of tiny particles. (It is a story which we shall tell in the following chapter.) The resulting wave/particle dilemma was resolved by Dirac in 1928 when he invented quantum field theory, a formalism which succeeds in combining waves and particles without a trace of paradox. Later developments in quantum electrodynamics (as the theory of the interaction of light and electrons is called) have led to the calculation of effects, such as the Lamb shift in hydrogen, which agree with experiment to the limits of available accuracy of a few parts per million. Can one doubt that such a tale is one of a tightening grasp of an actual reality? Of course there is an unusually strong element of corrigibility in this particular story. Quantum electrodynamics contains features completely contrary to the expectations which any nineteenth-century physicist could have entertained. Nevertheless there is also considerable continuity, with the concepts of wave and field playing vital roles throughout. The controlling element in this long development was not the ingenuity of men nor the pressure of society but the nature of the world as it was revealed to increasingly thorough investigation.

Considerations like these make scientists feel that they are right to take a philosophically realist view of the results of their researches; to suppose that they are finding out the way things are. When we are concerned with pre-quantum physics – with classical physics, as we say – that seems a particularly straightforward supposition. The analogy with the 'real' world of everyday experience is direct. In classical physics I can know both where an electron is and what it is doing. In more technical language, its position and momentum can both simultaneously be known. Such an object is not so very different from a table or a cow, concerning which I can have similar information of where they are and what they are doing. The classical electron can be conceived, so to speak, as just a midget brother of everyday things. Of course, philosophers can dispute the reality of the table and the cow too, but common sense is inclined to feel that that is a tiresomely perverse attitude to take to experience.

Heisenberg abolished such cosy picturability for quantum mechanical objects. His uncertainty principle . . . says that if I know where an electron is I have no idea of what it is doing and, conversely, if I know what it is doing I do not know where it is. The existence of such elusive objects clearly modifies our notion of reality. One of the perplexities about the interpretation of quantum mechanics is what, if any, meaning it attaches to the reality of something as protean as an electron. . . . [There] is in quantum theory a radical inability to pin things down which goes beyond the simple considerations outlined here.

The second perplexity relates to the act of measurement. It is notorious that there is an inescapable random element in quantum mechanical measurement. Suppose that I am supplied with a sequence of electrons which have been prepared in such a way that they are all in the same state of motion. Every minute, say, I am delivered one of these standard electrons. In classical physics, if I measured each electron's position as it was delivered they would all be found to be in the same place. This is because classically they have a well-defined location at a particular instant whose specification is part of what is involved in saying that they are in the same state of motion. Quantum mechanically, however, Heisenberg will not normally allow the electrons to have a well-determined position (it will usually have to be uncertain). This reflects itself in the fact that when I actually make a measurement I shall find the electron sometimes here and sometimes over there. If I make a large number of such measurements the theory enables me to calculate the proportion of times the electron will be found 'here' and the proportion of times it will be found 'there'. That is to say, the probability of its being found 'here' can be determined. However, I am not able to predict on any particular occasion whether the electron will turn out to be 'here' rather than 'there'. Quantum mechanics puts me in the position of a canny bookmaker who can calculate the odds that a horse may win in the course of the season but not in the position of Our Newmarket Correspondent who claims to be able to forecast the outcome of a particular race.

All that is rather peculiar but it can be digested and lived with. The source of the perplexity is more subtle. Consider what is involved in measuring the position of an electron. It requires the setting up of a chain of correlated consequences linking at one end the position of the microscopic electron and at the other end the registration of the result of that particular measurement. The latter can be thought of either as something like a pointer moving across a scale to a mark labelled 'here', or ultimately as a conscious observer looking at such a pointer and saying 'By Jove, the electron's "here" on this occasion.' Certainly I cannot perceive the electron directly. There has to be this chain of related consequence blowing-up the position of the microscopic electron into a macroscopically observable signal of its presence 'here'. The puzzle is, where along this chain does it get fixed that on this occasion the answer comes out 'here'? At one end of the chain there is a quantum mechanically uncertain electron; at the other a dependable pointer or equally reliable observer, neither of which exhibits any uncertainty in its behaviour. How do the two enmesh? . . . [We] shall find that it is a matter of perplexity where along the chain the fixity sets in which determines a particular result on a particular occasion. In fact there is a range of different suggestions, none of which appears free from difficulty. The discontinuity involved in the act of measurement is the one really novel feature which sets quantum mechanics apart from all the physics which preceded it.

In what lies ahead I shall try to explain more fully the nature of these problems and the variety of the answers which have been proposed to them. Before attempting that task a short historical excursus will be necessary to explain how it all came about.

Richard Proctor

A footnote in 'Astrology', from Myths and Marvels of Astronomy, 1903

There are few things more remarkable, or to reasoning minds more inexplicable, than the readiness with which men undertook in old times, and even now undertake, to interpret omens and assign prophetic significance to casual events. One can understand that foolish persons should believe in omens, and act upon the ideas suggested by their superstitions. The difficulty is to comprehend how these superstitions came into existence. For instance, who first conceived the idea that a particular line in the palm of the hand is the line of life; and what can possibly have suggested so absurd a notion? To whom did the thought first present itself that the pips on playing-cards are significant of future events; and why did he think so? How did the 'grounds' of a teacup come to acquire that deep significance which they now possess for Mrs Gamp and Betsy Prig? If the believers in these absurdities be asked *why* they believe, they answer readily enough either that they themselves or their friends have known remarkable fulfilments of the ominous indications of cards or tea-dregs, which must of necessity be the case where millions of forecasts are daily made by these instructive methods. But the persons who first invented those means of divination can have had no such reasons. They must have possessed imaginations of singular liveliness and not wanting in ingenuity. It is a pity that we know so little of them.

Charles E. Raven

An extract from *The Creator Spirit*, 1932

. . . Of course, if my 'self' is a mere bundle of instincts of known number and exact dimension, then let me tie the bundle up neatly and make the best of it; but if this elusive personality, with its queer and satisfying aspirations and relapses and struggles and touches of the eternal, is not just a machine with wheels that get out of order and a definite maximum horse-power, but a living thing indefinitely variable, constantly readjusting itself to circumstances, capable of incalculable achievement or of pathetic meanness, in some sense master of its fate; if its freedom is not an illusion, and its possibility of spiritual experience not a lie, then we must not allow ourselves to fall back into the old error of the mechanistic materialist.

Theodor Roszak

An extract from 'Technocracy's children', from *The Making of a Counter Culture*, 1973

By the technocracy, I mean that social form in which an industrial society reaches the peak of its organizational integration. It is the ideal men usually have in mind when they speak of modernizing, up-dating, rationalizing, planning. Drawing upon such unquestionable imperatives as the demand for efficiency, for social security, for large-scale co-ordination of men and resources, for ever higher levels of affluence and ever more impressive manifestations of collective human power, the technocracy works to knit together the anachronistic gaps and fissures of the industrial society. The meticulous systematization Adam Smith once celebrated in his well-known pin factory now extends to all areas of life, giving us human organization that matches the precision of our mechanistic organization. So we arrive at the era of social engineering in which entrepreneurial talent broadens its province to orchestrate the total human context which surrounds the industrial complex. Politics, education, leisure, entertainment, culture as a whole, the unconscious drives, and even . . . protest against the technocracy itself: all these become the subjects of purely technical scrutiny and of purely technical manipulation. The effort is to create a new social organism whose health depends upon its capacity to keep the technological heart beating regularly. . . .

In the technocracy, nothing is any longer small or simple or readily apparent to the non-technical man. Instead, the scale and intricacy of all human activities – political, economic, cultural – transcends the competence of the amateurish citizen and inexorably demands the attention of specially trained experts. Further, around this central core of experts who deal with large-scale public necessities, there grows up a circle of subsidiary experts who, battening on the general social prestige of technical skill in the technocracy, assume authoritative influence over even the most seemingly personal aspects of life: sexual behavior, child-rearing, mental health, recreation, etc. In the technocracy everything aspires to become purely technical, the subject of professional attention. The technocracy is therefore the regime of experts – or of those who can employ the experts. Among its key institutions we find the 'think-tank', in which is housed a multi-billion-dollar brainstorming industry that seeks to anticipate and integrate into the social planning quite simply everything on the scene. Thus, even before the general public has become fully aware of new developments, the technocracy has doped them out and laid its plans for adopting or rejecting, promoting or disparaging.

Within such a society, the citizen, confronted by bewildering bigness and complexity, finds it necessary to defer on all matters to those who know better. Indeed, it would be a violation of reason to do otherwise, since it is universally agreed that the prime goal of the society is to keep the productive apparatus turning over efficiently. In the absence of expertise, the great mechanism would surely bog down, leaving us in the midst of chaos and poverty.

Miriam Rothschild

'A liberating bolt from the blue', published in *The Scientist*, 1987

In November 1939, shortly after the outbreak of World War II, I was solving trematode life cycles in the Marine Biological Station at Plymouth. Before the fighting began I had received an impressive-looking form from the Royal Society asking for details of my qualifications, and announcing that as a scientist I was placed in a reserved occupation and could not volunteer for any form of national service should hostilities commence. I therefore continued with my work (I had already qualified as an air-raid warden) with a patriotic sideline: producing chicken feed made from seawood. I recall that the product I concocted gave me horrible hiccups.

The laboratory was totally unprepared for any form of aerial attack. The director, Stanley Kemp, adopted the view that (a) there would be no bombing from the air in this war, and (b) if there was aerial bombing Plymouth would escape it. Why? Because it was an open secret that the oil storage tanks had been emptied, the docks were insignificant and Plymouth was not on one of the main air lanes.

With great difficulty I persuaded the director, in my capacity as an air-raid warden, to allow me to provide the laboratory with fire-fighting equipment in the form of stirrup pumps and gas masks. It is difficult to believe, but this equipment, ridiculously inadequate as it was, was the only fire-fighting apparatus on the premises when the raids eventually began. No water had been stored in the massive tanks available, and of course the mains were soon put out of action. A year later, in the middle of one of the worst carpet-bombing raids of the war, a surrealistic scene occurred when the aquarium was partly demolished. A shell-shocked Indian student was standing knee-deep in water, with conger eels thrasing 'round him on the floor, aimlessly directing a thin dribble of water from one of my stirrup pumps at nothing in particular.

The first big air raid on Plymouth began at dusk, just before we separated to go to our respective lodgings. The laboratory had no proper air-raid shelter but the tunnel leading to the boat house gave some natural protection. We all spent the night there while sticks of bombs thrumped and crunched down on either side of us. Through a big open space in the roof of the tunnel we could see a purple square of sky, with silver pencils of the searchlights sweeping across it. Optimistically I wrote a long letter to an absent lover.

I asked Belinda Kemp, the director's daughter, who was about 17, if she was terribly frightened. 'No,' said Belinda thoughtfully. 'I have no imagination; I can't believe I won't be alive tomorrow.' Someone else remarked that it was immoral to hope that the bombs wouldn't fall on us since that meant they would fall on someone else. . . .

At 4 a.m. the raid was over and we lined up outside to watch, awestruck and appalled, while the oil tanks blazed. Of course they had not been emptied! Burning oil is terrifying. There is none of the spitting and crackling associated with an ordinary fire; it burns in silence except for a soft licking, bubbling murmur. There was a thick pall of smoke over everything. Not a word was spoken.

Eventually Dr Kemp kindly suggested that we go into his house, try to boil a kettle, and afterwards get some sleep in deck chairs. Following the cup of tea Mrs Kemp carefully divided the sexes and ushered us into separate rooms. I knew then that we were going to win the war.

When dawn came and there was light enough to see – for of course we had no artificial light working – I staggered to my room to assess the damage. An incredible sight met my eyes. The door had gone and the room appeared to be empty, except for a huge pile of tiny splinters of glass on the floor and, picking its way delicately among the debris, the sole survivor: my tame redshank.

Where were my notebooks and manuscripts? Where were the labelled drawings? Where were the cultures of intermediate hosts, the infected gobies, the hundreds of isolated infected snails? Where were the microscope, the Cambridge Rocker, the Camera Lucida, the watch glasses, the finger bowls, the tubes, the shelving, the jars? Gone. Seven years' work had vanished, pulverized with a ton of glass.

For three days I felt nothing except for a vague backache. I was stunned. A blank.

A German reconnaissance plane slipped in over the still-burning oil tanks, wheeled through the pall of smoke and vanished unmolested. Would we have another raid immediately? The fire surely provided a perfect visible target. Nothing happened.

Next morning I found that my redshank had died, possibly from delayed shock or some internal injury due to the blast. Had she died in great pain? I felt deeply disturbed as I saw her lying quietly among the slivers of broken glass – an unequivocal indictment of the human race. I grieved for her.

The following day I was seized by a sense of meaningless excitement and light-headedness. Without realizing it I had gradually become an appendage of my trematode life cycles. At the time I had no assistant, which meant that I could not afford a day's illness, let alone a vacation or a free weekend. There were all those intermediate hosts to tend and feed and rear; all those snails to nurture; scores of beautiful ephemeral cercariae to count, draw and describe; all those shells to measure; all those twinkling flame-cell patterns to unravel, and my flock of seagull hosts to coax from egg to adult. It had meant a remorseless 16-hour day. Why, I even saw cercariae in the clouds and flame cells beating in my dreams.

Now, all at once, I was free.

I packed my bags and left Plymouth, never to return as a research scientist. At the time I did not know that butterflies and fields of flowers were to be exchanged for free-swimming cercariae and the turbulent Atlantic Ocean. But temporarily, at least, the German Air Force had liberated me.

Richard Rowan-Robinson

Extracts from *Cosmic Landscape*, 1979

THE MEANING OF LANDSCAPE

Until it is tamed, landscape can seem harsh and hostile to human beings. Parts of the earth are permanently hostile by virtue of extremes of temperature – the polar wastes and the tropical deserts. Elsewhere people live with the shadow of earthquakes, volcanoes, floods, and hurricanes continually hanging over them. Even in more placid environments the farmer breaks his back over the soil and curses the fluctuations in rainfall which ruin his crops.

It is not surprising that so many images of nature, among both primitive and civilized peoples, are sombre and cataclysmic. Hence also the attraction of the closed garden, the secure, tamed piece of nature walled off from the world. Here man can gaze on nature in a calm and serene, but escapist, frame of mind.

In Western civilization it is during the Renaissance that we first find a more positive attitude to nature. The first person to climb a mountain to enjoy the view seems to have been Petrarch, and in his writings there is an all-embracing love of nature and the countryside: 'Would that you could know with what joy I wander free and alone among the mountains, forests and streams.' We find for the first time the concept of landscape, nature experienced as a whole by an individual: nature as a source of wonder and joy.

This all-embracing admiration of nature is at the heart of the scientific view of the world. It is not surprising that it should find its expression, however, in poetry rather than in the writings of the scientists themselves. The scientist is trying to comprehend the richness of nature and its phenomena rather than to convey this richness to others. Even so some scientific books, like Galileo's *Starry messenger* or Darwin's *Origin of species*, have become classics.

The earth offers such an immense range of landscape, from the dramatic wilds of desert and canyon to the patchwork of hedge and field of England, where even the remotest fell or moor bears traces of human occupation. But for me the essence of landscape is that there is a human being at the centre who experiences it. This is the sense in which I shall speak of the cosmic landscape. The landscape is an individual and personal experience. For example, each of us carries in our mind the landscape of childhood, those first recallable images of the outside world. And whether this landscape be harsh and alien or a secure garden it remains vivid and important throughout our lives.

THE COSMIC LANDSCAPE

When we look up at the night sky and see the moon, the planets, the stars, they do not seem part of our individual landscape. They are so remote and ethereal. We cannot travel to them, around them, we cannot look at them from different angles and from different distances as we can with the terrestrial landscape.

We see so little detail. To the naked eye only the moon shows any features. (To the inhabitants of ancient Teotihuacan in Mexico these features resembled a rabbit and they had an improbable story of how it got there.) The planets are slightly extended and they show some faint colour, as do the brighter stars in clear conditions. The changes that occur – the phases of the moon, the motion of the moon and planets through the stars, the nightly rotation of the sky, and the change from summer to winter constellations and back – all these have to be watched for regularly to see any pattern. Modern man is probably on average more ignorant of the night sky than the agricultural peoples of the remote past who used the aspect of the night sky to tell the season accurately. Indeed we have to marvel at the astronomical achievements of earlier, pre-telescopic cultures: the patient observations of the ancient Chaldeans in discovering the 18 years and 11 days cycle in the time that eclipses of the sun and moon occur, the care with which Chinese astronomers of the first millennium AD surveyed the stars to find changes in brightness or new visible objects, the precision with which the Maya were following the orbit of Venus at the time of the destruction of their books by the zealot Bishop Landa in 1562, the subtle discoveries of the medieval Arab astronomers, and the monumental work of the Greeks in the first 500 years BC.

There is a fascinating account by the Jesuit Lecomte of the working of the Chinese Astronomical Bureau as he found it in AD 1696:

> They still continue their observation. Five Mathematicians spend every night on the Tower in watching what passes overhead; one is gazing towards the Zenith, another to the East, a third to the West, the fourth turns his eyes Southwards, and a fifth Northwards, that nothing of what happens in the four Corners of the World may scape their diligent Observation.

It is very moving to think of the five mathematicians on the Tower, night after night for fifteen hundred years.

Of course an important change in our awareness of the cosmic landscape has happened in our times. Through the eye of the television camera we have ourselves seen the surfaces of the moon and Mars. I remember vividly the meeting of the Royal Astronomical Society in London in 1966 at which a slide was shown of the first picture of the moon's surface from NASA's *Surveyor I* camera. Suddenly there was a landscape with rocks and hills, rather like one of Leonardo's rugged backgrounds. Later there were the astronauts lumbering about in slow motion. And even those inarticulate pioneers conveyed one powerful image: how beautiful a haven the earth looked from that hostile terrain. The whole earth becomes the enclosed and secure garden, protected by the walls of its atmosphere and magnetic field. And the moon and Mars, scenes of so many wonderful fantasies of life on other worlds, are added to the list of magnificent but desolate products of attrition, the canyons and deserts. Places for the adventurous or foolhardy to visit in person but for all to visit in imagination.

How can this cosmic landscape of stars and those gigantic star systems, galaxies, become real for us? We have to travel in imagination to the universe's furthest shore. The cameras of the planetary probes take us only a minute step from our own soil. To comprehend the universe of the modern astronomer, we have to abandon our human scale of measurement. We have to leave solid ground behind, since lumps of rock like the earth are so rare. We have to become explorers without preconceptions. . . .

THE ALL-FREQUENCY LIGHT MACHINE

Let's imagine that we have a receiver that can tune to all possible wavebands – the all-frequency light machine. We transport it outside the earth's atmosphere and away from man-made interference, say to the moon. Each band is connected up to a suitable telescope which can instantaneously scan the whole sky. The results are fed to a special pair of spectacles so that you can see the sky as it appears to the telescope working in any particular band. What do you see as you tune across the wavebands? We shall leave explanations of the phenomena and the origins of these different radiations till later chapters. For the moment we concentrate on the appearance of the sky only.

We start with the visible band – that familiar sky of stars. We notice one thing immediately: the stars are not twinkling since there is no air to refract the starlight. We also see that the stars have different colours: most of the brighter ones are blue, but there is a good sprinkling of green, yellow, and red ones. The colours of the stars, which correspond to their different surface temperatures, can be seen with the naked eye from a good site on earth, for example a dry mountain site like Kitt Peak in Arizona. With a big telescope the stars appear as small pools of coloured light and from above the atmosphere they will appear as bright points of light.

That ribbon of light around the sky, the Milky Way, becomes resolved by our telescope into a giant disc of stars, our Galaxy. Those two fuzzy clouds of light noticed by Magellan in the southern sky are focused into other smaller star systems, galaxies orbiting our own Galaxy. Other more distant galaxies swim into view, some isolated, some in small groups, and some in huge swarms or clusters.

Now we shift the waveband past the blue end of the spectrum, past the violet and into a band of invisible radiation, the ultraviolet. The red, yellow, and green stars start to fade out, including the sun, while the blue ones become even brighter. Many of the galaxies, composed mainly of red stars, start to fade too. But in some we start to notice something interesting happening at the very centre of the galaxy, the nucleus. An intense point-like source of light starts to become more prominent. These are the 'active' galaxies, so called because of the evidence for explosions and other violent activity in their nuclei. The most intense of all, those in which this point-like source dominates the output even of all the stars in the galaxy, are called 'quasars', short for quasi-stellar radio source. These were discovered as star-like objects at the position of bright radio sources and only gradually realized to be the most distant and luminous objects in the universe. As we switch our waveband into the extreme ultraviolet a fog descends. Quasars, galaxies, the Milky Way, all fade from view and only the very nearest of the blue stars can be seen. The gas spread between the stars of the Milky Way has mopped up most of the light before it reaches us. This dark night of the far ultraviolet continues until we have increased the frequency (or shortened the wavelength) by another factor of ten. We have reached the 'soft' X-ray band and the fog turns to a light mist. Through the mist we see a different world, the world of gas at a temperature of a million degrees or more. The blue stars have almost faded out. Instead some pairs of stars that we had hardly noticed before blaze out. One of the stars is close to death. The Milky Way has returned, but instead of a broad river it has become a narrow stream. Into the 'hard' (higher frequency) X-ray band and the mist lifts completely. The Milky Way disappears as we explore the world of gas ten or a hundred million degrees centigrade or hotter. The dying stars in double-star systems are still there. So are the quasars and active galaxies, but normal galaxies are very weak. And a new kind of source is seen: the rich clusters of galaxies are bathed in diffuse light. We try to switch our all-frequency light machine to higher frequencies, the gamma-ray band, but the picture becomes very blurred and crude. We have reached the high-frequency limit of present-day astronomy.

What happens if we return to the visible band and switch in the opposite direction, towards lower frequencies and longer wavelengths? As we move from the red end of the visible band into the invisible 'near' (shorter wavelength) infrared the blue stars fade out and myriads of red stars swim into view, red giants of high mass and red dwarfs of low mass. Switching towards the middle infrared the stars start to fade out and cooler objects – planets, asteroids, and comets – become more noticeable. We are in the band dominated by dust warmed by stars – the interplanetary dust that causes the zodiacal light, cometary dust, and that precious speck of dust, earth itself. Certain of the red stars become prominent again because of the radiation from clouds of dust surrounding them, the cooled relics of materials thrown off by the stars. In some cases this dust cocoon is so thick that the star hardly shines through at all, so we notice them for the first time in this band. We move to the 'far' (longer wavelength) infrared, to the world of cold matter (−200°C or less). The stars have all vanished, even most of those shrouded in dust. Instead we see huge dimly glowing clouds of gas and dust spread round the Milky Way. In some, stars may just be beginning to form out of the gas. The nucleus of our own Galaxy, in the direction of the constellation of Sagittarius, the Archer, stands out because of the great concentration of stars there, buried in thick clouds of dust. The nuclei of some other galaxies are even more dramatic, in some cases because of a burst of newly formed stars, in others because an active nucleus is hidden within the dust.

Between the longest far-infrared and the shortest radio wavelengths, in the 'microwave' band, we become aware of a faint sea of light all round the sky, the 'cosmic microwave background' radiation, relic of the 'Big-Bang' origin of the universe over ten thousand million years ago. The gas clouds of the Milky Way begin to be illuminated in the vicinity of bright blue stars that have recently formed out of the gas of the clouds. We notice the quasars and active galaxies becoming bright again. And at certain characteristic wavelengths the world of interstellar molecules lights up.

Into the radio band and the Milky Way has spread out into a bloated avenue of light, much brighter towards the direction of the constellation of Sagittarius than in the opposite direction, the constellation of Taurus the Bull. The plane of the Milky Way is still picked out with radiation from hot gas near young blue stars. The rest of the sky is rather uniformly covered with fainter sources, mostly active galaxies and quasars. Other near-by normal galaxies seem very weak by comparison.

As we move to lover radio frequencies we notice a smooth haze of background radiation coming from the whole sky. By comparison the Milky Way, which has spread out to cover almost the whole sky, starts to fade and the clouds of hot gas, which were so prominent at high radio frequencies, suddenly vanish and turn into dark patches. The same clouds of hot gas are now absorbing the radio waves from behind them. Thus high and low radio frequencies are like positive and negative copies of a photograph, with bright sources on a dark sky at high frequencies and dark holes on a bright sky at low frequencies. Many of the sources spread over the sky have faded too and the sky is becoming more sparsely populated than at any time since we passed the gamma-ray band, apart from the fog of the extreme ultraviolet. Before we can decide at what frequencies these sources too fade, our receiver has begun to crackle and hiss. We have again reached the limit of the known world.

But what exactly is this all-frequency light machine? It is none other than we ourselves, and these telescopes are our eyes. New eyes that are the product of our cultural evolution. . . .

WHAT WOULD IT BE LIKE TO ENCOUNTER THEM?

Supposing the optimistic calculations are correct, what are we likely to encounter out there? It seems exceedingly unlikely that our first encounter would be with people at the same stage of development as ourselves. Our forty years with radio telescopes follow three thousand million years of evolution. Even if others developed in an almost identical way they are unlikely to reach this stage at the same date or even within a thousand or a million years. There are stars a thousand million years older than the sun just as suitable for the evolution of life. When we think of the voyages of discovery of the European Renaissance, the navigators and travellers found peoples at vastly different stages of development, even though these peoples all had a common origin. We who have only just begun to see the cosmic landscape are likely to be the undeveloped primitive in any interstellar conversation. It is also inevitable that with their thousands or millions of years of technology they will discover us before we discover them. Our earliest radio broadcasts at frequencies high enough to penetrate the ionosphere may even now be being detected by creatures with unimaginably advanced technology, who are now deciding whether to ignore us, colonize us or look after us. It seems to me that if such creatures exist relatively near earth we are in a very vulnerable state. Perhaps our days as an independent planet are numbered. In sending out signals deliberately beamed at near-by stars we are giving a hostage to fortune. Would the Indians of the Americas have sent such a signal to Europe before 1492 if they had been able to? We have to hope that civilizations much more advanced than ourselves are less aggressive, less selfish, and less violent than we are at the present era. Perhaps they have to be to survive the discovery of nuclear physics. But perhaps they have to be selfish to survive.

If science and technology continue to develop on earth for thousands or millions of years, it is clear that we shall be taking a great interest in the question of other habitable planets, so that we can emigrate either because of some kind of cosmic calamity here (change of climate, reversal of magnetic field, near-by supernova) or just for fun. Others may take a similar interest in earth now and might find it very underpopulated and rich in natural resources compared with their own planet.

Human beings have a natural hunger to know if there is other intelligent life in the cosmic landscape, even if only so that we can wave to each other across the gulf of interstellar space and feel less alone in the immensity of space and time. Yet the most likely circumstances of such an encounter, with creatures far more developed than ourselves, seem fraught with insecurity and danger. An encounter with creatures like ourselves does not seem an attractive idea either, with both planets changing their inflated military machines over from 'defence' weapons directed against other nations to 'defence' weapons directed against other solar systems.

The ideal encounter would be with some civilization slightly more advanced than ourselves, mature, gentle, with no interest in colonizing us, who could help us sort out our problems. This hunger for a cosmic parent or psychiatrist will get worse if things keep on going so badly on earth. But is it likely that such people exist? Why are we here after all? There seems to be no reason, except as a product of the programme of certain molecules to replicate themselves. The molecules made us, but they did not necessarily have in mind coping with the nuclear age. Was there no advantage to the molecules in making us more mature?

Yet we do not have to be ruled by the molecules, for our cultural evolution enables us to transcend their limitations. And it is our culture that would help us bear slavery if that were to become our fate. Like the ancient Greeks under Roman rule we would intellectually and spiritually colonize our conquerers, transforming them into Byzantines.

In the face of these apocalyptic visions of the future and of the vague menace of civilizations a thousand or a thousand million years more advanced than ourselves, we turn again to earth, to the secure garden of the terrestrial landscape. We do not do so in quite the same frame of mind as the medieval artists. Our *Vision of Judgement* is not peopled by demons like the landscape of Hieronymous Bosch, though the creatures that alarm us might look just as fantastical. And the garden we return to is not quite the tame, ornamental one of the *Lady with the Unicorn* tapestries in Paris. A better parallel is with Leonardo da Vinci, with his interest in astronomy, his scientific desire to describe and understand nature, and his obsession with the idea of the Deluge, a product of the apocalyptic speculations current around 1500.

We have seen that we can experience the cosmic landscape in all its aspects by an imaginative extension of our senses. Our bodies can feel the infrared radiation of the sun and our skin responds to the ultraviolet by getting tanned. The X- and gamma rays of the cosmic landscape, mainly from the surface of our radioactive earth, have had a profound effect on our evolution from primitive forms of life through the mutations they cause. The human body radiates mostly in the infrared band, but it also radiates weakly at microwave and even at radio wavelengths. We can think of the astronomer's diverse telescopes, and the strange detectors with which he records radio, microwave, infrared, ultraviolet, and X- and gamma radiation, as an extension of our eyes. In the first chapter we called this the all-frequency light machine.

With these new eyes we saw a series of landscapes, or rather views of the same landscape. In the visible band we saw beyond the landscape of planets and stars to giant star systems like the Milky Way, the galaxies, spread out in every direction as far as the eye could see. In the radio we saw along the Milky Way clouds of hot ionized gas around hot stars and by contrast very cold clouds of gas where stars have not yet formed. We saw the astonishingly regular flashing of the pulsars, hinting at the dramatic death of stars as supernovae and neutron stars. Beyond the Milky Way we saw the violent activity of radio galaxies and we discovered the awe-inspiring power of the quasars, power equivalent to a thousand Milky Ways from a volume no bigger than the solar system. In the ultraviolet band we were surprised by the million-degree gas of the sun's corona. The hot, young luminous stars and the hot, dying white dwarfs blazed out. We saw that many galaxies have miniature quasars in their very heart. With X-rays the dead stars were brought to life, white dwarfs, neutron stars, and the sinister black holes, lit up by gas from their living companions pouring onto them. We saw the full drama of a solar flare and the other active sources, quasars and Seyfert galaxies, raged too. The giant clusters of galaxies glowed with gas stripped from their member galaxies. In the infrared band we saw the planets and asteroids of the solar system at their strongest. Stars shrouded in dust swam into view, both stars being born and stars close to death in their red-giant phase. We saw the clouds where stars are forming and the cold clouds whose turn is yet to come. We saw galaxies with quasars in their centre shrouded in dust. In the microwave band we saw the cosmic whisper that was all that remained of the Big Bang and we travelled back along the photon's path to the fireball itself. The world of molecules lit up and we found everywhere the organic molecules that are the first steps towards life itself.

And this landscape is our own personal landscape. The atoms of our body were in the cold clouds and in the hot centres of stars. Earlier they floated on the fireball of light that was the Big Bang. Yet looming larger still in the foreground is the landscape of earth, more important to us than all the rest of the universe. We have to care for this precious and rare environment that made us and that we are now beginning to be capable of destroying. There may be no other earths in the Milky Way. We probably do not have any contemporaries out there. Earth and its life are a marvel and mankind, with its culture, its sciences, and its arts, is the greatest marvel of all. The barren landscapes of the moon and Mars appear more familiar and real to us than the remoter parts of the cosmic landscape because they seem to have been seen through our own eyes and have acquired a human dimension. I have to admit that it is harder to feel the same way about the landscape seen with the telescope, the landscape of the invisible wavelengths, though the whole aim of this book has been to bring that landscape closer.

Is this cosmic landscape the only one we could have lived in? Are there other possible universes, with other histories, in which we could have existed? Our existence seems bound up with so many aspects of the cosmic landscape, perhaps all aspects if we could only see the connection. This point of view has been elevated by some cosmologists to a law, the anthropic principle: things are as they are because if they were not we would not be here. Several strange coincidences in the ratios between physical constants can be explained in this way, for all the different physical forces, gravitational, electromagnetic, nuclear, have played their part in our evolution. If gravitation had been a bit stronger or a bit weaker, stars like the sun or planets like the earth might not have formed and we would not be here. We see that the main scientific technique of the past four hundred years since the time of Galileo, that of examining a particular physical process in isolation from all others, is not suitable when we come to look at the universe as a whole. There is only one universe, the parts of it are not separate, and it is not a laboratory with many different experiments bubbling away. It is a complete and interconnected unity and we are an integral part of it.

Travelling through the cosmic landscape is a wonderful experience and we do not have to let the vast distances and times intimidate us. It is absurd to be depressed by the thought of the sun's extinction 5 thousand million years hence, or the universe's possible collapse back to infinite destiny in a hundred thousand million years. It is we, a mere bag of molecules scurrying round on a grain of interstellar dust, who have made this voyage and given the universe a consciousness of itself. The universe might have existed without us, but it would not have become a landscape.

Oliver Sacks

'The President's speech', from *The Man who Mistook his Wife for a Hat*

What was going on? A roar of laughter from the aphasia ward, just as the President's speech was starting, and the patients had all been so eager to hear the President speak.

There he was, the old charmer, the actor with his practiced rhetoric, his histrionics, his emotional appeal – and all the patients were convulsed with laughter. Well, not all: some looked bewildered, some looked outraged, one or two looked apprehensive, but most looked amused. The President is generally thought to be a moving speaker – but he was moving them, apparently, mainly to laughter. What could they be thinking? Were they failing to understand him? Or did they, perhaps, understand him all too well?

It was often said of these patients, who though intelligent had the severest receptive aphasia, rendering them incapable of understanding words as such, that they nonetheless understood most of what was said to them. Their friends, their relatives, the nurses who knew them well, could hardly believe, sometimes, that they *were* aphasic. This was because, when addressed naturally, they grasped some or most of the meaning. And one does speak 'naturally', naturally.

Thus to demonstrate their aphasia, one had to go to extraordinary lengths, as a neurologist, to speak and behave unnaturally, to remove all the extraverbal cues – tone of voice, intonation, suggestive emphasis or inflection, as well as all visual cues (one's expressions, one's gestures, one's entire, largely unconscious, personal repertoire and posture). One had to remove all of this (which might involve total concealment of one's person, and total depersonalization of one's voice, even to use a computerized voice synthesizer) in order to reduce speech to pure words, speech totally devoid of what Frege called 'tone-color' (*Klangenfarben*) or 'evocation'. With the most sensitive patients, it was only with such a grossly artificial, mechanical speech, somewhat like that of the computers in *Star Trek*, that one could be wholly sure of their aphasia.

Why all this? Because speech – natural speech – does *not* consist of words alone, or (as the English neurologist Hughlings Jackson thought) of 'propositions' alone. It consists of *utterance* – an uttering forth of one's whole meaning with one's whole being – the understanding of which involves infinitely more than mere word recognition. This was the clue to aphasiacs' understanding, even when they might be wholly uncomprehending of words as such. For though the words, the verbal constructions, per se, might convey nothing, spoken language is normally suffused with 'tone', embedded in an expressiveness that transcends the verbal. It is precisely this expressiveness, so deep, so various, so complex, so subtle, that is perfectly preserved in aphasia, though understanding of words be destroyed. Preserved, and often more: preternaturally enhanced.

This too becomes clear, often in the most striking or comic or dramatic way, to all those who work or live closely with aphasiacs: their families or friends or nurses or doctors. At first, perhaps, we see nothing much the matter; and then we see that there has been a great change, almost an inversion, in their understanding of speech. Something has gone, been devastated, it is true – but something has come in its stead, has been immensely enhanced, so that, at least with emotionally laden utterance, the meaning may be fully grasped even when every word is missed. This, in our species *Homo loquens*, seems almost an inversion of the usual order of things; an inversion, and perhaps a reversion too, to something more primitive and elemental. And this perhaps is why Hughlings Jackson compared aphasiacs to dogs (a comparison that might outrage both), though when he did this he was chiefly thinking of their linguistic incompetence, rather than their remarkable and almost infallible sensitivity to 'tone' and feeling. Henry Head, more sensitive in this regard, speaks of 'feeling-tone' in his treatise *Aphasia* (1926), and stresses how it is preserved, and often enhanced, in aphasiacs.

Thus the feeling I sometimes have – which all of us who work closely with aphasiacs have – that one cannot lie to an aphasiac. He cannot grasp your words, and so cannot be deceived by them; but what he grasps he grasps with infallible precision, namely the *expression* that goes with the words, that total, spontaneous, involuntary expressiveness which can never be simulated or faked, as words alone can, all too easily.

We recognize this with dogs, and often use them for this purpose – to pick up falsehood, or malice, or equivocal intentions, to tell us who can be trusted, who is integral, who makes sense – when we, so susceptible to words, cannot trust our own instincts.

And what dogs can do here, aphasiacs do too, and at a human and immeasurably superior level. 'One can lie with the mouth,' Nietzsche writes, 'but with the accompanying grimace one nevertheless tells the truth.' To such a grimace, to any falsity or impropriety in bodily appearance or posture, aphasiacs are preternaturally sensitive. And if they cannot see one – this is especially true of our blind aphasiacs – they have an infallible ear for every vocal nuance, the tone, the rhythm, the cadences, the music, the subtlest modulations, inflections, intonations, which can give, or remove, verisimilitude from a man's voice.

In this, then, lies their power of understanding – understanding, without words, what is authentic or inauthentic. Thus it was the grimaces, the histrionisms, the gestures – and, above all, the tones and cadences of the President's voice – that rang false for these wordless but immensely sensitive patients. It was to these (for them) most glaring, even grotesque, incongruities and improprieties that my aphasiac patients responded, undeceived and undeceivable by words.

This is why they laughed at the President's speech.

If one cannot lie to an aphasiac, in view of his special sensitivity to expression and 'tone', how is it, we might ask, with patients – if there are such – who *lack* any sense of expression and 'tone', while preserving, unchanged, their comprehension for words: patients of an exactly opposite kind? We have a number of such patients, also on the aphasia ward, although technically they do not have aphasia but, instead, a form of agnosia, in particular a so-called tonal agnosia. For such patients, typically, the expressive qualities of voices disappear – their tone, their timbre, their feeling, their entire character – while words (and grammatical constructions) are perfectly understood. Such tonal agnosias (or 'atonias') are associated with disorders of the right temporal lobe of the brain, whereas the aphasias go with disorders of the left temporal lobe.

Among the patients with tonal agnosia on our aphasia ward who also listened to the President's speech was Edith D., with a glioma in her right temporal lobe. A former English teacher and poet of some repute, with an exceptional feeling for language, and strong powers of analysis and expression, Edith was able to articulate the opposite situation – how the President's speech sounded to someone with tonal agnosia. Edith could no longer tell if a voice was angry, cheerful, sad – whatever. Since voices now lacked expression, she had to look at people's faces, their postures and movements when they talked, and found herself doing so with a care, an intensity I had never seen her do before. But this, it so happened, was also limited, because she had a malignant glaucoma and was rapidly losing her sight too.

What she then found she had to do was to pay extreme attention to the exactness of words and word use, and to insist that those around her did just the same. She could less and less follow loose speech or slang – speech of an allusive or emotional kind – and more and more required of her interlocutors that they speak *prose*, 'proper words in proper places'. Prose might, she found, compensate, in some degree, for lack of perceived tone or feeling.

In this way she was able to preserve, even enhance, the use of 'expressive' speech (to use Frege's distinction), in which meaning is largely given by the apt choice and reference of words, despite being more and more lost with 'evocative' speech, where meaning is largely given in the use and sense of tone.

Edith also listened, stony-faced, to the President's speech, bringing to it a strange mixture of enhanced and defective perceptions – precisely the opposite mixture from those of our aphasiacs. It did not move her – no speech now moved her – and all that was evocative, genuine or false, completely passed her by. Deprived of emotional reaction, was she then transported or taken in? By no means. 'He is not cogent', she said. 'He does not speak good prose. His word use is improper. Either he is brain-damaged, or he has something to conceal.' Thus the President's speech did not work for Edith either, owing to her enhanced sense of formal language use, propriety as prose, any more than it worked for our aphasiacs, with their word-deafness but enhanced sense of tone.

Here then was the paradoxical possibility raised by the President's speech. That a good many normal people, aided, doubtless, by their wish to be fooled, were indeed well and truly fooled ('*Populus vult decipi, ergo decipiatur*'). And so cunningly was deceptive word use combined with deceptive tone, that it was the brain-damaged who remained undeceived.

Charles Sherrington

Extracts from Man on his Nature, 1940

The eye's parts are objects familiar even apart from technical knowledge and have evident fitness for their special purposes. The likeness to an optical camera is plain beyond seeking. If a craftsman sought to construct an optical camera, let us say for photography, he would turn for his materials to wood and metal and glass. He would not expect to have to provide the actual motor power adjusting the focal length or the size of the aperture admitting light. He would leave the motor power out. If told to relinquish wood and metal and glass and to use instead some albumen, salt and water, he certainly would not proceed even to begin. Yet this is what that little pin's-head bud of multiplying cells, the starting embryo, proceeds to do. And in a number of weeks it will have all ready. I call it a bud, but it is a system separate from that of its parent, although feeding itself on juices from its mother. And the eye it is going to make will be made out of those juices. Its whole self is at its setting out not one ten-thousandth part the size of the eye-ball it sets about to produce. Indeed it will make two eyeballs built and finished to one standard so that the mind can read their two pictures together as one. The magic in those juices goes by the chemical names, protein, sugar, fat, salts, water. Of them 80% is water.

Water is a great menstruum of 'life'. It makes life possible. It was part of the plot by which our planet engendered life. Every egg-cell is mostly water, and water is its first habitat. Water it turns to endless purposes; mechanical support and bed for its membranous sheets as they form and shape and fold. The early embryo is largely membranes. Here a particular piece grows fast because its cells do so. There is bulges or dips, to do this or that or simply to find room for itself. At some other centre of special activity the sheet will thicken. Again at some other place it will thin and form a hole. That is how the mouth, which at first leads nowhere, presently opens into the stomach. In the doing of all this, water is a main means.

The eye-ball is a little camera. Its smallness is part of its perfection. A spheroid camera. There are not many anatomical organs where exact shape counts for so much as with the eye. Light which will enter the eye will traverse a lens placed in the right position there. *Will* traverse; all this making of the eye which *will* see in the light is carried out in the dark. It is a preparing in darkness for use in light. The lens required is biconvex and to be shaped truly enough to focus its pencil of light at the particular distance of the sheet of photosensitive cells at the back, the retina. The biconvex lens is made of cells, like those of the skin but modified to be glass-clear. It is delicately slung with accurate centring across the path of the light which *will* in due time some months later enter the eye. In front of it a circular screen controls, like the iris-stop of a camera or microscope, the width of the beam and is adjustable, so that in a poor light more is taken for the image. In a microscope, or photographic camera, this adjustment is made by the observer working the instrument. In the eye this adjustment is automatic, worked by the image itself!

The lens and screen cut the chamber of the eye into a front half and a back half, both filled with clear humour, practically water, kept under a certain pressure maintaining the eye-ball's right shape. The front chamber is completed by a layer of skin specialized to be glass-clear, and free from blood-vessels which if present would with their blood throw shadows within the eye. This living glass-clear sheet is covered with a layer of tear-water constantly renewed. This tear-water has the special chemical power of killing germs which might inflame the eye. This glass-clear bit of skin has only one of the four-fold set of the skin-senses; its touch is always 'pain', for it should *not* be touched. The skin above and below this window grows into movable flaps, dry outside like ordinary skin, but moist inside so as to wipe the window clean every minute or so from any specks of dust, by painting over it fresh tear-water.

We must not dwell on points of detail; our time precludes them, remarkable though they are. The light-sensitive screen at back is the key-structure. It registers a continually changing picture. It receives, takes and records a moving picture life-long without change of 'plate', through every waking day. It signals its shifting exposures to the brain.

This camera also focuses itself automatically, according to the distance of the picture interesting it. It makes its lens 'stronger' or 'weaker' as required. This camera also turns itself in the direction of the view required. It is moreover contrived as though with forethought of self-preservation. Should danger threaten, in a moment its skin shutters close protecting its transparent window. And the whole structure, with its prescience and all its efficiency, is produced by and out of specks of granular slime arranging themselves as of their own accord in sheets and layers, and acting seemingly on an agreed plan. That done, and their organ complete, they abide by what they have accomplished. They lapse into relative quietude and change no more. It all sounds an unskilful overstated tale which challenges belief. But to faithful observation so it is. There is more yet.

The little hollow bladder of the embryo-brain, narrowing itself at two points so as to be triple, thrusts from its foremost chamber to either side a hollow bud. This bud pushes toward the overlying skin. That skin, as though it knew and sympathized, then dips down forming a cuplike hollow to meet the hollow brain-stalk growing outward. They meet. The round end of the hollow brain-bud dimples inward and becomes a cup. Concurrently, the ingrowth from the skin nips itself free from its original skin. It rounds itself into a hollow ball, lying in the mouth of the brain-cup. Of this stalked cup, the optic cup, the stalk becomes in a few weeks a cable of a million nerve-fibres connecting the nerve-cells within the eye-ball itself with the brain. The optic cup, at first just a two-deep layer of somewhat simple-looking cells, multiplies its layers at the bottom of the cup where, when light enters the eye – which will not be for some weeks yet – the photo-image will in due course lie. There the layer becomes a fourfold layer of great complexity. It is strictly speaking a piece of the brain lying within the eye-ball. Indeed the whole brain itself, traced back to its embryonic beginning, is found to be all of a piece with the primordial skin – a primordial gesture as if to inculcate Aristotle's maxim about sense and mind.

The deepest cells at the bottom of the cup become a photo-sensitive layer – the sensitive film of the camera. If light is to act on the retina – and it is from the retina that light's visual effect is known to start – it must be absorbed there. In the retina a delicate purplish pigment absorbs incident light and is bleached by it, giving a light-picture. The photo-chemical effect generates nerve-currents running to the brain.

The nerve-lines connecting the photo-sensitive layer with the brain are not simple. They are in series of relays. It is the primitive cells of the optic cup, they and their progeny, which become in a few weeks these relays resembling a little brain, and each and all so shaped and connected as to transmit duly to the right points of the brain itself each light-picture momentarily formed and 'taken'. On the sense-cell layer the 'image' has, picture-like, two dimensions. These space-relations 'reappear' in the mind; hence we may think their data in the picture are in some way preserved in the electrical patterning of the resultant disturbance in the brain. But reminding us that the step from electrical disturbance in the brain to the mental experience is the mystery it is, the mind adds the third dimension when interpreting the two-dimensional picture! Also it adds colour; in short it makes a three-dimensional visual scene out of an electrical disturbance. . . .

Suppose we choose the hour of deep sleep. Then only in some sparse and out of the way places are nodes flashing and trains of light-points running. Such places indicate local activity still in progress. At one such place we can watch the behaviour of a group of lights perhaps a myriad strong. They are pursuing a mystic and recurrent manoeuvre as if of some incantational dance. They are superintending the beating of the heart and the state of the arteries so that while we sleep the circulation of the blood is what it should be. The great knotted headpiece of the whole sleeping system lies for the most part dark, and quite especially so the roof-brain. Occasionally at places in it lighted points flash or move but soon subside. Such lighted points and moving trains of lights are mainly far in the outskirts, and wink slowly and travel slowly. At intervals even a gush of sparks wells up and sends a train down the spinal cord, only to fail to arouse it. Where however the stalk joins the headpiece, there goes forward in a limited field a remarkable display. A dense constellation of some thousands of nodal points bursts out every few seconds into a short phase of rhythmical flashing. At first a few lights, then more, increasing in rate and number with a deliberate crescendo to a climax, then to decline and die away. After due pause the efflorescence is repeated. With each such rhythmic outburst goes a discharge of trains of travelling lights along the stalk and out of it altogether into a number of nerve-branches. What is this doing? It manages the taking of our breath the while we sleep.

Should we continue to watch the scheme we should observe after a time an impressive change which suddenly accrues. In the great head-end which has been mostly darkness spring up myriads of twinkling stationary lights and myriads of trains of moving lights of many different directions. It is as though activity from one of those local places which continued restless in the darkened main-mass suddenly spread far and wide and invaded all. The great topmost sheet of the mass, that where hardly a light had twinkled or moved, becomes now a sparkling field of rhythmic flashing points with trains of travelling sparks hurrying hither and thither. The brain is waking and with it the mind is returning. It is as if the Milky Way entered upon some cosmic dance. Swiftly the head-mass becomes an enchanted loom where millions of flashing shuttles weave a dissolving pattern, always a meaningful pattern though never an abiding one; a shifting harmony of subpatterns. Now as the waking body rouses, subpatterns of this great harmony of activity stretch down into the unlit tracks of the stalk-piece of the scheme. Strings of flashing and travelling sparks engage the lengths of it. This means that the body is up and rises to meet its waking day.

Dissolving pattern after dissolving pattern will, the long day through, without remission melt into and succeed each other in this scheme by which for the moment we figure the brain and spinal cord. Especially, and with complexity incredible, in that part which we were thinking of, the roof-brain. Only after day is done will it quiet down, lapse half-way to extinction, and fall again asleep. Then at last, so far at least as the roof-brain, motor acts cease. The brain is released from the waking day and marshals its factors for its motor acts no more. . . .

John Maynard Smith

'The definition of life', chapter 1 of *The Problems of Biology*, 1986

With the advent of space travel, the question has become more pressing: how shall we decide whether something is alive? If there are, on other worlds, objects as large and complex as elephants and oak trees, most of us are reasonably confident that we could recognize them as living. We are less confident about objects as small and, relatively, as simple as viruses and bacteria, but the difficulty arises because of the inadequacy of our sense organs rather than of our concepts. If we had eyes which magnified like an electron microscope, we would probably recognize a bacterium as alive.

Can we make more precise the criteria we would use? I suggest that there are two relevant properties:

(i) Although the forms of living organisms remain constant, the atoms and molecules of which they are composed are constantly changing; in other words, they have a 'metabolism'.

(ii) The parts of organisms have 'functions'; that is, the parts contribute to the survival and reproduction of the whole. Thus your legs are for walking with, your heart is for pumping blood round your body, and the feathery seed-heads of dandelions help to disperse the seed.

The major part of biology is concerned with these two properties of organisms: biochemistry and physiology with the former, genetics and evolution theory with the latter. . . .

Organisms are not the only objects which maintain a fixed form despite a continuous change in their constituent molecules. Consider two non-living examples. When you pull the plug out of a basin, the escaping water will form a rotating mass surrounding a narrow column of air; that is, it will form a vortex. If the level in the basin is kept constant by adding water, the form of the vortex remains constant, while the molecules of water pass through it and down the plughole. To take a second example, if you light a Bunsen burner, the flame has a fixed form, with different colours in different regions. Again, so long as there is a flow of gas, the form remains constant.

Both flame and vortex are examples of 'dissipative structures'. They are structures whose maintenance requires a continuous input of energy, and whose effect is to dissipate that energy. In the vortex, the energy is the potential energy of the water, which is dissipated as the water falls. In the flame, the energy from chemical reactions is dissipated as heat. As soon as the supply of energy stops, the form disappears. (In this respect, the conditions which preserve the structure are the exact opposite of those which preserve the form of a building or a snowflake, which last better in the absence of an input of energy.)

To what extent are living organisms dissipative structures? It is not true that all the atoms in an organism, as in a vortex or a flame, are in a state of flux. Many large molecules (proteins and nucleic acids) remain unchanged throughout the adult life of an organism, and are composed of the same atoms at the end of adult life as at the beginning. The extent to which this is true varies greatly between organisms (it is more true of adult insects than of vertebrates), between parts of organisms (it is more true of your brain than your intestine), and between kinds of molecules. In some quite complex organisms (e.g. some insects), this permanent molecular structure is so arranged that the animal can be freeze-dried, so that almost all chemical reactions cease, and can then be brought to life again.

Nevertheless, it is true that the maintenance of the living state requires a constant flow of energy through the system. A freeze-dried insect is not alive: it was alive, and may be alive again in the future. Energy must be supplied either in the form of suitable chemical compounds or as sunlight, and in either case atoms are continuously entering and leaving the structure of the organism. The difference is that the structure of even the simplest organisms is enormously more complex than that of a vortex or a flame, and that the flow of energy is controlled. This control depends on the continued existence of the large molecules mentioned in the last paragraph, and of larger structures made from them. The way these controls operate is the main subject-matter of biochemistry and physiology.

The use of the word 'control' brings us sharply up against the second characteristic of life: that the parts of living organisms contribute to the survival of the whole. Living things are not the only objects that have 'controls'. One of the simplest controls is the governor of a steam engine, whose rotating arms swing out further as the engine goes faster, and, by reducing the supply of steam, control the engine speed.

It was this parallel between machines and organisms which provided one of the central arguments of Natural Theology – that is, of that branch of theology which seeks to give reasons for a belief in God derived from a consideration of the natural world. William Paley, the most famous of the Victorian natural theologians, argued as follows. If we look at a piece of machinery such as a watch, it is apparent that its structure has a purpose – telling the time. From this we deduce that the watch must have been designed, and hence that it had a designer. Since we see that living organisms have the same characteristics – that their parts have functions or purposes – we can again deduce that they were designed. Hence we can deduce the existence of God, the designer.

Before discussing how modern biologists interpret functions, one difference between the views of theologians and biologists needs comment. We agree that the heart has a function: we differ only as to how it acquired that function. But do elephants or oak trees have functions? My own view is that they do not: it is as senseless to ask what an elephant is for as it is to ask what an electron is for. But for some theologians, the world was created for men to live in, and so animals have a purpose: to serve man. To the question 'what are seals for?' one theologian answered that they are there to give polar bears something to eat, because otherwise the bears would come south and eat us. Apart from the infinite regress implicit in this reply ('What are polar bears for?' and so on), few biologists would accept the anthropocentric assumption. . . .

Yet I agree with Paley that the heart has a function, even if we disagree about the elephant. But biologists no longer accept the argument from design, because they have an alternative explanation of how the parts of organisms come to have functions. This alternative is Darwin's theory of evolution by natural selection.

Darwin's theory can be formulated as follows. Given a population of entities having the properties of multiplication, variation, and heredity, and given that some of the variation affects the success of these entities in surviving and multiplying, then that population will evolve: that is, the nature of its constituent entities will change in time. Of the three essential properties, multiplication means that one entity can give rise to two, and variation that not all entities are identical. Heredity means that like begets like. . . . the essential point is that when multiplication occurs, kind A usually gives rise to A, kind B to B, and so on. If heredity were exact, evolutionary change would eventually slow down and stop; continuing evolution requires that heredity is inexact, so that new variants arise from time to time.

Given a population of this kind, we can say rather more than that evolution will occur; we can say something about the kinds of changes which will occur. The population will come to consist of entities with properties which help them to survive and reproduce. Note that this does not give us a fully predictive theory of evolution, because the course of evolution will also depend on the repertoire of variation which arises. But we do expect organisms which have evolved by natural selection to have organs which help them to survive and reproduce: that is, we expect them to have parts with 'functions'.

So formulated, the theory of evolution is not a falsifiable scientific theory in the sense required by Karl Popper, who demands of a theory that it should rule out certain kinds of events which, if they do in fact occur, can be used to show that the theory is false. As formulated in the last two paragraphs, Darwin's theory is not falsifiable. It is a statement of a logical deduction: if certain things are true, then evolution follows. A scientific theory must say something about the world, and not merely about logical necessity. Darwinism as a testable scientific theory can take various forms. I will give it first in the form in which Darwin himself proposed it, and then in the 'neo-Darwinist' form in which most biologists hold it today.

For Darwin, the theory had two components. The first was the proposition that all living things on earth are descended, with modification, from one or a few simple original forms. He did not claim to account for the origin of those first living things. This part of Darwin's theory, sometimes referred to as the 'fact' of evolution, is almost universally accepted by biologists today. The second component of the theory was that the major (but not the only) cause of evolutionary change was natural selection. As a subsidiary cause of evolution, Darwin accepted 'the effects of use and disuse'; to over-simplify, he thought that organisms acquired during their lifetimes changed characteristics (called by biologists 'characters' or 'traits'), which they would then pass on to their offspring.

The claim of descent with modification is clearly falsifiable; as J. B. S. Haldane remarked, a single fossil rabbit in Cambrian rocks would be sufficient, because the first fossil mammals are found in rocks some 400 million years later than the Cambrian. The theory that natural selection is one cause of evolution is not so easily falsifiable. It would be falsified if one could show that organisms lack one of the three necessary properties of multiplication, variation, and heredity, but I know of no one who has ever attempted to do so.

Thus I think it may be an error to try to force Darwin too strictly into the Popperian mould; after all, Popper's philosophy was derived from a study of physics. Darwin's claim that evolution has happened certainly falls within Popper's criterion. His proposed mechanism of natural selection, however, does not, because he put it forward as only one among several possible mechanisms and not as the only one, so that any observation which could not be explained by selection could always be explained in some other way. As I shall explain later, modern evolution theory is closer to meeting Popper's criterion, because it is no longer so ready to accept alternative explanations of adaptation.

Darwin's position is best summarized as follows. He started with the observation that organisms are adapted to survive in particular environments, and that they have parts which function so as to ensure that survival. That is, his starting-point was that any adequate theory of evolution must explain adaptation. In his theory of natural selection, he pointed out that organisms do in fact have the properties of multiplication, variation, and heredity, and that it is a necessary consequence of this that they should become adapted for survival. That is, he showed that there is a necessary connection between one set of observations, concerning reproduction, and another set, concerning adaptation.

Our main difference from Darwin lies in our possession of a theory of heredity, outlined in the next chapter. This has led us to develop Darwin's ideas in two main ways:

(i) As discussed in detail later, we no longer think that changes that occur during an individual lifetime as a consequence of the use or disuse of organs alter the nature of the children produced by that individual. Hence we no longer regard the 'inheritance of acquired characters' as a significant cause of evolution. In this sense, we are more Darwinian than Darwin, since we regard natural selection as the only relevant mechanism leading to the evolution of adaptation, instead of merely the most important one.

(ii) We are more interested in other possible causes of evolutionary change. There are various possibilities. First, there are changes (variously referred to as 'genetic drift' and as 'non-Darwinian evolution') which occur because the traits concerned have so little effect on survival and reproduction that the changes are essentially random. Second, the entities which are subject to evolution by natural selection may not be individual organisms, but either larger entities (populations) or smaller ones (genes, or groups of genes). Third, changes may occur because information can be transmitted between generations not only genetically, but also culturally.

These are changes of emphasis rather than substance. The central point remains that Darwin provided a theory which predicts that organisms should have parts adapted to ensure their survival and reproduction. This has led to the suggestion that life should be defined by the possession of those properties which are needed to ensure evolution by natural selection. That is, entities with the properties of multiplication, variation, and heredity are alive, and entities lacking one or more of those properties are not.

There is much to be said for this definition. Consider, for example, the question of whether fire is alive. One answer might be that it is not, because although it has a structure which remains constant despite a continuous change of substance, that structure is too simple to qualify as living. But that leaves us with the unilluminating question of how complicated an object must be to qualify as living. It is more illuminating to say that fire is not alive because, although it has the properties of multiplication (one fire can light another) and variation, it lacks heredity. Thus, although fires vary in size, colour, and temperature, their nature at any instant depends on immediate circumstances – the supply of fuel, how much wind there is, and so on – and not at all on whether they were started with a match or a cigarette-lighter. Because, with fires, like does not beget like, fires cannot evolve by natural selection, and so do not become complex, and do not acquire organs to keep them going.

From this preliminary survey, two distinct pictures of living organisms have emerged. One is of a population of entities which, because they possess a hereditary mechanism, will evolve adaptations for survival. The other is of a complex structure which is maintained by the energy flowing through it. . . . Perhaps the hardest part of biology is to see how the two pictures fit together. Most questions in biology require at least two answers, one appropriate to each of the two pictures. Suppose we ask, 'why does the heart beat?' One answer will be concerned with how differences in heart shape are inherited, and how they influence the survival of their possessors, and will lead to the conclusion that the function of the heartbeat is to pump blood round the body. The other will be concerned with the rhythmical properties of heart muscle, and with the nerves supplying the heart, and will lead to an account of heart physiology. Rather unhappily, these two kinds of answers have been called, respectively, the 'functional' and 'causal' explanations. The words are unhappy, because the 'functional' explanation is also a causal one, but on a longer time-scale. It is perhaps better to refer to the two kinds of explanation as the ultimate and the proximal causes of the heart beating. The important thing to understand, however, is that the two explanations are not alternative and mutually exclusive; both can be true, and they can illuminate one another.

George Steiner

'Has truth a future?', Bronowski Memorial Lecture, 1978

Jokes and anecdotes are the radio-active tracers in human history. By their light we can make out cardinal moments in the pulse and growth of feeling. Antique mockery has it that Thales fell down a well. The mathematical wizard who, legend insists, foretold the solar eclipse in 585 BC, could not make out the ground in front of his own feet. When Syracuse was captured in 212 BC, invading soldiers broke into Archimedes' garden. His mechanical devices had kept the assailants at bay. Now they were out for blood. Bent over a problem in the geometry of conic sections, Archimedes did not hear his killers coming. He perished, as it were, in a fit of abstraction.

The first tale is one of shrewd derision, the second of wondering discomfort. Both are ancient, and tell of an overwhelming turn in human affairs. They declare the irony, the amazement, the stab of the uncanny which men and women experienced when there arose among them the seeker after abstract truth, the hunter after disinterested knowledge, the pure mathematician, the logician and ponderer on metaphysics. It is in the abstraction, in the absolute disinterestedness, that lay the unsettling marvel of the thing. There had long been surveyors who applied trigonometry to farmlands and building sites, astrologers seeking to harness the future by means of spherical geometry, artisans of analytic argument who brought logic to bear on matters of law and administration. *Applied* science is as old as man's notice of sunrise. But Thales stumbling into the well and Archimedes enraptured unto death by an equation, represent something overwhelmingly different.

First and foremost, their addition is with the abstract, the inapplicable, the sovereignly useless. Application, where it comes at all, comes *after*. It is the bonus, the impurity of condescension that may come of truth. The purer the object of inquiry, the more remote from everyday instrumentality or profit, the finer, the more passionate the inquiry, the more intense 'the danse of the mind' (a Platonic image, which renaissance art will turn into literal motion). To solve an equation, to map the conjunction of celestial bodies millennia after the life of the astronomer, to find algebraic or verbal expression for the flux of invisible atoms, to debate of ideal forms and the nature of immortality, is to pursue the truth for its own sake. Not – and this is the inhuman crux – or not in the first place, for the sake of justice on earth, of a better life for the common man, of wealth or political power. But in a passionate autism, for no reason but its own, its beauty or that sharp edge of beauty which we call difficulty. Thales does not look at the ground before him, Archimedes does not hear his executioners.

The disinterested pursuit of abstract truths of a scientific and philosophic nature, the very notion of the truth as falsifiable, as subject to experimental trial or logical constraint, is *not* a universal. This seems a startling paradox. Almost unconsciously, we associate the word 'truth' with something 'out there', eternal, serenely independent of historical, local circumstance. But the concept of abstract truth and of the pursuit of abstract truth, be it at the · cost of good sense, of social utility, of economic benefit, of life itself, springs up in one place and one place only, and at one specific and singular time in our history. The hunters after abstract truth are, to the best of our knowledge, men of Ionia and of the Greek colonies in southern Italy. Their wondrous infatuation arises at some point between the seventh and fifth centuries BC. It does *not* arise in any other civilisation or place or time. In varying degrees all human communities have their *shamans*, their measurers, their healers. These are applied scientists. The man prepared to risk destitution or ridicule in order to resolve a quadratic equation or a paradox in logic, is an Ionian or south Italian Greek in the age of Pythagoras and Anaximander, of Heraclitus and Anaxagoras. Those who today pursue abstract truths in the sciences, in mathematics, in philosophy, are heirs to his hunger.

Why did this speculative lust originate in the Mediterranean world at the time of the pre-Socratics? This is an exceedingly difficult question, and one that may lead to ground which is, by current standards of liberal humaneness, dangerous and even repellent. Montesquieu speculated that climate might have been a vital factor. The pursuit of mathematical and philosophic truths, a delight and virtuosity in abstract argument, might have been particularly fostered by the out-of-door character of Ionian and Sicilian life, by the bracing exactitude of Mediterranean light and the prodigal exhibit of the heavens. Modern sociologists of culture have pointed to the possibility that the diet of the early Greek city-states was superior, notably in protein, to that of surrounding cultures; that mental powers are, in a literal sense, nourished. The societies in question were male-dominated to an extreme degree. For reasons which we are only beginning to apprehend dimly, the contribution of women to pure mathematics, to metaphysical speculation and to that allegory of mathematics which is music, has been very small. It was so from the outset. And there is the pivotal point of leisure. We are looking at a *polis* in which men were given time, space and authority to think abstractly, to think in defiance of utility not only by servitude of women but by that of slaves. Thales had no need to do his own cooking; it is improbable that it was Archimedes himself who ran that famous bath.

But whatever the complex causes for the insinuation into the spirit of western man of the drug of truth, the quality of the obsession is clear from the start. The search for truth is predatory. It is a literal hunt, a conquest. There is that exemplary instant in Book IV of *The Republic*, when Socrates and his companions in discourse corner an abstract truth. They halloo, like hunters who have unearthed and run down their quarry.

That cry rang down the ages. In the renaissance, in the seventeenth century, the model of pursuit remained mathematical. Man after Pythagoras had discovered in himself a strange power: that of unfolding, from a set of abstract definitions and axioms, an unbounded, forward-leaping chain of new propositions, theorems, algorithms which might, indeed, possess practical bearing, which might number the universe, but whose essential meaning and elegance were wholly autonomous. Man, in his infirmity of biological being, could, by virtue of energies and exact dreams still not understood, formulate mathematical problems whose solution might lie a thousand years ahead and, what is even more uncanny, he could discriminate between those problems to which there would, one day, be a solution, and those which would remain undecideable – a vaulting across time and the unknown which appears to be as unique, as central to the genius of man, as is the invention of melody.

Even where it focused on moral, on aesthetic, on metaphysical concerns, the main tradition of western philosophic thought kept in steady view the ideals of mathematical rigour and purity set forth by Plato. For both Descartes and Kant the problem was precisely that of understanding how the human mind could operate mathematically, how it could manoeuvre with immaculate abstruseness, and whether the structure of reality at large matched the lineaments of deduction and of logic drawn by man. At the same time, the formidable development of the applied sciences and of technology brought on a morality of the fact. To get facts right, to accumulate them and put them in order became an essential part of the pursuit of the truth. Linnaeus's classification of flora, the great Encyclopédie, are mastering nets thrown over the resistant manifold of the world.

But whether in the abstract sphere or in the empirical, in the disinterested or the utilitarian, one imperative prevailed: that human dignity, excellence, prosperity and happiness can only benefit from the hunt after truth, from the discovery of new truths. The Gospel promise that 'the truth shall make you free' was equally self-evident, equally crucial, to the humanists of the renaissance and the natural philosophers of the enlightenment. 'The truth alone is beautiful, only truth is to be loved,' said Montaigne. This was the commonplace of pride and of hope from Marsilio

Ficino to Thomas Jefferson. This postulate had manifest political consequences. It was the implicit theology of liberalism. It is because the truth will prevail, that freedom of inquiry, of public proposal and debate are, as Milton had proclaimed, the indispensable but also the most efficacious means of human progress. And it is because he assists most visibly in the discovery and dissemination of the truth, that the scholar, the scientist, the publicist is a benefactor of mankind whose often eccentric, idiosyncratic and even costly labours should be underwritten by the body politick. From the renaissance to Condorcet, from Condorcet to John Stuart Mill, it is held as axiomatic that the forward march of individual · men and of society is inseparably inwoven with the pursuit of truth and with the application of this pursuit to the humanities, to the arts, to the sciences and to technology.

But central and eloquent as it is, this position, with its live roots in the discovery of abstract, disinterested thought in pre-Socratic Hellas, has not gone unchallenged. The laughter against Thales, the vengeance on Archimedes, have their own strong history.

The first line of attack is that of the mystic and irrationalist. It professes the part of Asia in the western spirit. It was present already in the Greek mystery cults and in the characteristic Hellenic fascination with Egypt and Persia. Here the vision of truth is not analytic, but transcendent. It springs not from deduction but illumination. The criteria of authentic discovery are not experimental control and refutability, but an immediate and visionary light that seizes upon the soul. The illuminist tradition, whether it be in St John's revelation of the Word that is beyond logic, in the exultant play with contradictions of the Pietists, or, as it sometimes seems, at the far edge of Wittgenstein's model of language and the world, assigns a different order of truth to the sciences, to historical documentation, to the social fact. Almost invariably, this assignment is also a derogation. The truths of the mathematician, of the physicist, of the logician or historian, are felt to be of a lower type. They belong to the temporal and the contingent. They lack the seven seals of annunciation and the certitude of rapture. The given, the bestowed quality of this certitude is a cardinal point. The truths of the mystic are not hammered out of the stubborn matter of existence. They are experienced, they are, in the image of St John of the Cross, that supreme disciplinarian of ecstasy, 'suffered in the inmost'. Almost by definition, this sufference of truth entails a posture which is contemplative, which sees action, notably of the questing mind, as impatient and predatory.

A second line of attack is launched from a foundation of dogma, of systematic religious and supernatural revelation. Ultimate verities are the monopoly of the deity or of the divine pantheon. 'I am the truth' says Christ, a sentence which is untranslatable into the semantics of the Greek philosophers and mathematicians. What is offered to mortal man is that part of revealed truth which accords with his limited estate and capacities. Again, as in the tradition of outright mysticism, the kinds of truth aimed at by experiment, by axiomatic deduction, by scientific analysis, are assigned to an inferior order. They come of secular hazard. They are not eternal. They do not possess the authority of the divine statement.

But this statement differs from mystical illumination in that it will often aspire to a logic, to a 'rationality', as soberly-construed as any in the natural and philosophic sciences. Hence the ambiguities and bitterness of the struggles between dogmatic christianity on the one hand and Galilean or Darwinian science on the other. The struggle is not one between unreason and rationality, between apocalypse and logic. It is a contest, the grimmer for its intimacy, between competing analyses of evidence and causal relation, between rival claims to logic and to proof. The essential tenor of this duel is often misunderstood. Christianity did not fear that it could not find a place within its own constructs for Galilean astronomy or the evolution of species. Rightly, it saw in such scientific hypotheses an inherent instability, a built-in mechanism of self-challenge and revision. Thus the sciences expose to chance and seemingly violent renovation those very concepts of truth and of proof to which a revealed religion must attribute everlastingness and perfect internal coherence. At stake in the trial of Galileo and the condemnation of Darwinism was not this or that set of facts, but the notion of the fact itself and of the relationship of the fact to the divinely underwritten perception of truth.

There is a third line of attack on the pursuit of abstract, objective truth as the Platonic huntsmen understood it. This line draws obviously on both the mystical and the theological traditions, but it gives to them its own psychological twist. I would call it the romantic existential polemic. We find it exemplified in such figures as Blake, Wordsworth, Kierkegaard, Dostoevski, Nietzsche and Tolstoy. The values of ecstatic self-realisation, of moral goodness, of simplicity, even of absurdity, are set high above those of mathematical-scientific truth. In Blake's celebrated critique of Newton, there is alarm at the possibility that the mathematical, analytic and taxonomic world-view of post-Galilean science will quench the demon of sensory delight. There is also the shrewd premonition that abstract science and the ideals of rationality must bring with them the technological revolution. The prismatic decomposition of the rainbow in Newton's optics, and the effacement of this same rainbow behind factory steam are, as Blake saw, rigorously connected. In both, the truth is tyrannical. The Wordsworthian attack is that of spontaneous feeling against abstruse mentalism, of body against cerebration, of intuitive simplicity against mandarin difficulty. In his season of revolutionary hopes, Wordsworth had believed in a fusion between the arts and the sciences, he had believed that the 'impassioned countenance of the sciences' would irradiate the spirit of the poet. But political disappointment came, and with it the conviction that 'one impulse from a vernal wood' outweighed the dusty fruit of libraries and laboratories. When Kierkegaard opposes the 'abyss of the absurd' to the pretences of man's reason and logic, he is advancing not only a fierce satire on the naive self-congratulations of nineteenth-century positivism and Hegelian historicism. He is attempting to recall men and women to the mystery of their incompletion, to the radical absurdity of their everyday condition when that condition entails an understanding neither of birth nor · of death. Dostoevski's 'man from underground' proclaims that two plus two equals five. He does so in order to reassert the scandal and wonder of human freedom as against the dictates of mathematical postulates and the prohibitions of logic. In Nietzsche, the criterion is one of instinctive wisdom over sterile deduction, of the vertigo of self which makes it impossible to know the dancer from the dance (Yeats' image), as contrasted with the impersonality, with the self-suppression of the pursuit of objective, scientific aims. To Tolstoy the sagacity of the unlettered child, the innocent clairvoyance of the old peasant, came to weigh infinitely more – in terms of ethics *and* of mental sanity – than the jargon of the philosopher and the hollow promises of the scientist or engineer.

The fourth line of attack is in many respects the subtlest, the most penetrating. It has its distant origins in Greek scepticism. We hear it in Pascal's remark that there is one truth this side of the Pyrenees but another on the Spanish side. Or in Lenin's dictum: 'Do not ask whether a thing is true or not; ask only, *true for whom?*' The most recent form of this attack is to be found in the social theory of the Frankfurt School, in Horkheimer's, Adorno's and Marcuse's critiques of the enlightenment. The argument is this: objectivity, scientific laws, the concept of truth and falsehood, logic itself, are neither eternal nor neutral. They express and enact the world-view, the economic purposes and power-structures, the class interests of that dominant élite which began in the slave-economy of ancient Greece and reached its full flowering in the mercantile-technological imperialism of western capitalist societies. The notion of abstract truth, of the ineluctable fact, are instruments not of disinterestedness but of the class struggle. What hungry man feeds on the theory of relativity? 'Truth' is a complex variable depending on the social context. There is no objective history, but a history of oppressor and oppressed. Logic is a weapon of the literate against the intuitive-sensory modes of speech and sentiment among the masses. The enshrinement and teaching of scientific laws, be they Newtonian, Malthusian or Darwinian, reflects a conscious means of spiritual and material control over society. 'The truth, as taught by your masters, shall make you slaves.'

Today, these four lines of attack – the mystical, the religious-dogmatic, the romantic-existentialist and the relativist, or dialectical – are conjoined. In diverse shadings and proportions, they make up the ideological stock of what are called 'counter-culture movements'. They animate the pastoral hopes of proponents of 'an alternative technology' (where Blake invoked 'the holiness of the minute particular', we proclaim that 'small is beautiful'). The widespread rejection by the young of numeracy, of the sciences, of technology, are inspired by these critiques. The children of Krishna tango down our sullen streets; undergraduates in the university of Bacon, Newton and Darwin crowd to hear the latest guru out of the Californian orient; there are at the most recent count three times as many registered astrologers in western Europe and the United States as there are members of professional associations of physics and of chemistry. In the White House, a successor to Jefferson has just expressed the belief that he has seen an Unidentified Flying Object. All about us are the drop-outs of reason. But, I repeat: these radical-pastoralist and visionary lines of doubt or rejection are ancient. They were active in the world, indeed in the divided spirit, of Pythagoras.

But now, and for the first time in man's history, there is a new and total challenge to the ideal of the hunt after pure, abstract truth. For the first time, one can conceive of a fundamental incongruence, of a fundamental coming out of phase, between the pursuit of truth and equally demanding ideals of social justice or, even more centrally, between truth and survival. The stakes in the obsessive race after discovery are no longer a broken leg for Thales or Archimedes' mortal inadvertence. They might be, possibly they already are, the continued existence of the individual and of society as we have known them. It was as if certain orders of truth had grown irreconcilable to man. Let me cite three examples in ascending categories of immediacy. The first is, almost absurdly, remote.

The history of man's analysis of the thermodynamics of the universe is, unquestionably, one of the most absorbing in that of the species as a whole. Speculations on the death of the cosmos by ice or by fire are as ancient as human wonder and the analogous induction of the inevitable death of the individual. In the thought of the Stoics and of the Gnostics, there are intimations of a slow demise of the world, through the fatigue inherent in corrupt and hybrid matter. But it was not until 1824, that Carnot's mathematical formulation of entropy gave to these archetypal conjectures a precise, a determinist validation. Developed by Clapeyron, by Clausius, by Kelvin. Carnot's demonstration of the necessary loss of energy in any and all mechanical systems, was extended to the thermodynamics and kinetics of the universe as a whole. In a celebrated letter, Keats exclaims that he would value immensely the knowledge of whether Shakespeare sat or stood, was tranquil or febrile, when he composed this or that supreme passage. Would it not be as arresting to know of Clausius's stance and state of mind when, in 1865, he set down on paper the phrase 'the heat-death of the universe'? The algebra of the case is not very sophisticated, but the proof is stringent. And the conclusions that follow are, as Bertrand Russell put them, comprehensive:

> The second law of thermodynamics makes it scarcely possible to doubt that the universe is running down, and that ultimately nothing of the slightest interest will be possible anywhere. Of course, it is open to us to say that when that time comes God will wind up the machinery again: but if we do say this, we can base our assertion only upon faith, not upon one shred of scientific evidence. So far as scientific evidence goes, the universe has crawled by slow stages to a somewhat pitiful result on this earth, and is going to crawl by still more pitiful stages to the condition of universal death.

You will say, rightly, that this heat-death is so distant in time that our determination of its inevitability remains a pure game of the intellect, that the truth thus captured cannot press on man or society. But localise entropy, apply it to the mere solar system, the time-order of whose disappearance in the exploding radiance of the aged sun has been calculated, and the walls of the absolute come closer. I recall the shock I experienced as an adolescent when George Gamow's brilliant exposition of the death of the sun made plain to me that nothing would survive of our present world – not a line of Plato, not a verse of Shakespeare, not a note of Mozart. Still, such annihilations are immensely remote. They do, as the German poet Paul Celan expressed it, 'lie north of the future'.

So let us come down to earth with a second example. The suspicion that aggression and violence, ranging from a private brawl to full-scale war, are an integral element in human nature, is as old as Homer. The mutual destruction which men have visited upon one another since the beginning of recorded history would not be a chain of accidents, an infantile disorder in the race, a murderous frivolity or, as Clausewitz contended, a more or less calculated extension of the economic and foreign policies between rationally-governed communities. Homicidal warfare would be the collective expression of instinctual needs and characteristics in the human primate. Even as a muscle injects acids into the bloodstream or retracts, first into pain and then into atrophy, when it is unused, so there are in man primary impulses of assault, of sacrificial risk, which must be enacted if the human psyche is to remain in creative poise. No political enlightenment, no reform of communal institutions would touch the roots of desire which compel men to identify, to invent if necessary, to seek out, their enemies.

During prolonged periods of peace (and these have been rare in history), the muscles of aggression, of disciplined hatred, would stiffen into acute discomfort. Men would find themselves breathing hard and short in what a troubadour singer and swordsman called 'the stench of peace'. They would search, with mounting stress, for war-surrogates, as do the tribes in the interior of New Guinea when they meet, at prescribed times and places, to wage ritualised and often sanguinary mock-battles.

Is this analysis true? Are there in the human animal needs of aggression and of fatal risk which must be satisfied if the nervous system is to preserve its intricate dynamics? Much social evidence seems to point in this · direction. The urban violence in western streets, the apparently motiveless lust for mayhem in the football hooligan and city vandal, can be read all too graphically as substitutes of war. One need only glance at the bookstalls, cinemas, television-screen of Britain today to recognise an almost uncontrolled nostalgia for the bitter splendour that was world war one, for the finest hour that was world war two. The nostalgia aches. And rarely before has a society been more conscious of *ennui*, of the thin greyness of peace. Or contrast the near-total eclipse of interest, of recollection, in regard to the landings on the moon, with the fascination, the metaphoric presentness of the events, memories, images of recent wars as they exercise our dreams.

But the evidence remains uncertain. We do not know whether there is such a thing as an instinct to aggression. We do not know whether violence is a necessary factor in the economy of the organism. But the question facing us today is simply this: ought we to find out? Should we pursue research into the nature of aggression, into the anatomy of murderousness, at the risk of a positive finding? Should we attempt to establish, uncompromisingly, whether Hegel was right when he put forward the supposition that human identity fully realises itself only in the mastery over, in the destruction of another identity? If it were so, what then? Given the nuclear armistice, what options have we, what modes of symbolic or ritual war-games are left to us? Or will our survival come to depend on the enforcement of an agreed fiction, to the effect that man is, as Rousseau would have him be, essentially benevolent, that wars are a black error of the past? And if Homer did know better, when he saw in battle the high noon and garland of man's energies, where shall this knowing end?

But again you may say that this dilemma is not immediate, that the physiological and behavioural sciences relevant to the issue are not, can never be, of imperative exactness. There could always be room for doubt and adjournment. I am not certain. The precarious well-being of the industrial west now seems to depend on a demonstrably lunatic armaments race, on the manufacture of overkill with built-in obsolescence. To many men, caught in the abjection of declining living standards, the miasma of peace seems more suffocating than the bracing air of war. Deep-seated mechanisms of unrest are at work. But perhaps they can be cauterised. So let us come nearer still.

Debate rages over the entire, vast field of human heredity and the radical techniques of the new biology which are loosely called 'genetic engineering'. Should cloning be allowed; should research into and experimentation with recombinant DNA be pursued or stopped; what of the creation of organic, self-replicating matter *in vitro*? Surrounding, penetrating these issues in complex, often distorting ways, is the more general question of the relations between heredity and environment, between genetic factors and historical-social circumstance. In the fate of the individual, how much weight are we to assign to inheritance, how much to nurture and milieu? The highly technical character of the evidence and techniques involved, notably in molecular biology, in distributional statistics, in the very definition of what constitutes a genetic code, a gene-pool and the osmotic interface between the genetic and the environmental, makes it exceedingly difficult for the layman to grasp the arguments on either side, let alone to arrive at his own verdict. What is clear is this: the questions being asked, the possibilities being envisaged, the choices being laid before us are, quite literally, matters of life and death for the social system as we have known it.

Examples are at once obvious and numbing. Pre-determination of the sex of the foetus, the ability for man to choose whether he wishes a male or a female offspring, is now in prospect. Are we to make use of this utterly undreamt-of option, which, given linear programming of sufficient sophistication, could become a political-military instrument of the most extreme consequence and coercion? Might this sorcery of decision prove invaluable in the event of a nuclear catastrophe? What of the creation, or to use more cautious terms, of the molecular reproduction or combination of life in the laboratory? Again, one understands, the requisite techniques are not very far off. What could be the potential effects on society? What might be the effects – and this is a far more subtle question – on the psychic equilibrium of the human species?

But even these challenges, overpowering as they are, seem momentarily academic and tractable compared to the debate over heredity and environment. If it was to emerge, from rigorous research, that innate factors in the human species greatly outweigh in importance those of the environment, that these inherited, innate factors are diversely distributed among ethnic groups, and that ameliorations in the environment can only marginally affect the unfolding of human talents and achievements – what then? Suppose we came to understand that it is inheritance which enables one quite limited ethnic strain to leap over jumps of more than seven feet almost as a matter of routine, whereas another ethnic group will continue to produce the theorems of higher mathematics and of physics – and that to train the first in the abstract sciences and the second in athletics would not, statistically, affect the outcome – what then? The answer that we must contrive a world in which high-jumping is as richly rewarded as intellectual genius is facile enough. In many respects we already have such a world. But to say that the one is as important as the other is pure cant. We have known otherwise, since Thales stumbled and Archimedes was absent-minded. And those who howl in the public halls in which these issues are to be raised are afraid. They may be right.

Should genetic inquiry go forward, whatever the social, the human consequence? Are there orders of abstract truth irreconcilable to human justice? These are not remote conundrums. The issue lies before us, here and now.

One answer is obvious: to say 'hold, enough'. It can be urged, persuasively, on economic grounds. In a world of hunger and destitution, what rational, what humane apologia can there be for the investment of literally hundreds of millions of dollars, pounds or roubles in the study of subatomic particles? When the means are lacking merely to repair our hospitals and schools, let alone purge our megalopolis of its slums, what possible defence can there be for the expenditure of enormous capital on radio-telescopes, on space-probes, on satellites designed to map the distribution of x-ray sources in the galaxy? All of which are not only fantastically costly to obtain, but incomprehensible to the vast majority of mankind.

But the ethical, political polemic is more drastic. It concedes that the insights being striven for by the molecular biologist, by the manipulator of recombinant DNA, by the comparative geneticist, may be of the highest degree of intellectual fascination, that there are in sight of the hunt verities which could alter our conception of man and of his evolution. But it insists that these verities are of a kind which society cannot handle. More penetratingly than even the most hazardous of radiations, certain orders of truth would infect the marrow of politics and would poison beyond cure the already tense relations between social classes and ethnic communities. In short, there are doors immediately in front of current research which are marked 'too dangerous to open', which would, if we were to force them, open on chaos and inhumanity.

There are scientists who advocate such inhibition in either an absolute or qualified vein. There are research-teams that have broken up over the use of cloning and the investigation of inherited characteristics. There are younger physicists and biochemists in the forefront of the ecological insurgence. There is defensiveness and doubt in the house of the exact sciences.

Drawing on the traditions of attack which I have already mentioned, an alternative vision now declares itself. Science can be, it must be, made 'socially responsible'. It must be scaled down to human appliance. The common cold cries out for the attention of genius. The Aristotelian and Cartesian commitment to theoretic analysis has led western man into his present estate of bewildered ferocity. Let us put the questing, autistic brain to pasture while instinct plays. Not *homo cogitans*, as philosophers and mathematician would have him, but *homo ludens*. Not the costly, élitist chase after the esoteric and possibly destructive fact, but the arcadian pilgrimage towards self and community. If there can be an alternative technology of low-energy consumption, of recycling, of ecological conservation, why not an alternative logic, an alternative style of cerebration? Before he was a hunter and killer, man was a gatherer of berries just this side of Eden.

It won't work. First, for reasons which are pragmatic and technical. 'Big science' is no accidental outcome of megalomania. It is inherent in the internal development of physics, of chemistry, of molecular biology, of astronomy and astrophysics. Collaborative – if often bitterly competitive – research and large-scale instrumentation have sprung inevitably out of the nature of the problems to be solved. Where such problems, moreover, touch on the dwindling energy resources of the planet, only the most massive scale of experiment and analysis gives promise of progress. The notion that science can again grow small, that the radar-dish can become the astrolabe, is, I think, wholly unrealistic. Added to this is the plain fact that even the most abstruse of research, into the 'strangeness' of the muon or into the mathematics of the genetic code, *may*, even suddenly, prove to have practical application (Rutherford's confident error with respect to the total non-applicability of nuclear physics is one we shall not have the good fortune to repeat). Thus, it is highly improbable that rival nation-states will inhibit or humanise the scientific research on which their future grandeur or security might depend.

But the real reason why we shall not return to innocence lies much deeper. The obsession with objective and abstract truth is imprinted on the western mind. Certainly the right circumstances, climate, economic margins, professional potentialities are needed to trigger the impulse after truth into creative being. But the daemon is there, and very close, I suspect, to the core of our identity and culture. We shall continue to ask wherever the question may lead, however dangerous the answer might prove to be. Cut off public funds, censor the open debate of explosive issues, break up the great laboratories, threaten the pure scientist with ostracism and moral anathema: the search will go on. Somewhere, at some moment, a man alone, a group of men addicted to the drug of absolute thought, will be seeking to create organic tissue, to determine the nature of heredity, to produce the cloud-chamber trail of quarks. Not for renown, not for the benefit of the human species, not in the name of social justice or profit, but because of a drive stronger than love, stronger even than hatred, *which is to be interested in something*. For its own enigmatic sake. Because it is there.

It may be that western man has been, since the Greek *polis*, and that he is still, a predator, that there is in his pursuit of the truth an imperialism analogous to that of his colonial conquests and economic exploitations. It may be that he has been programmed to surmount obstacles, whatever the cost. But there is another way of looking at it. The disinterestedness of mathematical, scientific, philosophic speculation, the conspicuous consumption of economic resources and personal existence on behalf of abstruse, useless truth, could be our singular dignity. The capacity to fall down wells while meditating on the motion of the stars or of sacrificing one's life through the supreme inattention of concentrated analytic thought, may be the best excuse there is for man. Given the choice between personal safety and social progress on the one hand and the solution to Fermat's last theorem or a proof for the Goldbach conjecture on prime numbers on the other, there will always be, there must always be, those who opt for the solution.

Archimedes knew this and acted on his knowing. What is new is only this: the possibility that the truths which lie ahead of us are not merely extraneous to man, or incomprehensible to all but a handful of men – these negations have long been with us – but that there may be between man's survival and certain truths, between man's hopes of justice and decency and certain categories of truth, a fundamental incompatibility. It is now conceivable that the truth lies in wait for man, as it did for Oedipus, the solver of riddles, where the three ways meet. It is conceivable that the rage for insight, the hunter's cry, which infected the western mind almost three thousand years ago, is leading us into ambush.

The man whom we have gathered to honour and remember this evening was a mathematician and a physical chemist, a virtuoso of speculative thought and a far-seeing technologist. He was a fine poet and a lover of the arts. Above all else, he was a being endowed with a veritable genius for the communication of his discoveries and passions to others. For millions who saw and who read *The Ascent of Man*, Jacob Bronowski embodied, as no one else in our time has, the search for truth, but also the conviction that this search must be made understandable, must be made 'bracing' – a word beloved of Blake who was Bronowski's favourite poet – to ordinary men and women. There was in him an elfin pleasure in mathematical difficulty, in the leap of the tensed intellect across the laziness of illusion or conventional 'good sense', but also an abounding sympathy for those less gifted than he was. Bronowski would not condescend. Hence his triumph, his unforgettable presentness of heart and mind, on the disenchanted medium of television.

Few men of science and of thought were as well equipped as was Bronowski to pose precisely and attempt an answer to the dilemmas which I have tried to lay before you. The search for a humanistic, philosophically-informed biology, for a genuine 'science of life', which occupied so much of his concern in his later years, springs directly from the questions raised tonight. The man to give this lecture was, of course, Bronowski himself.

Had he done so, he would, I suppose, have argued that the trap can be avoided, that there is always a fourth road which Oedipus the truth-hunter was too arrogant, too impatient to perceive. If forced to the wall, Bronowski would have said, indeed he did say, that no truth must ever be suppressed, that 'it doesn't matter whether you're talking about bombs or the intelligence quotients of one race as against another . . . if a man is a scientist, like me, he'll always say "Publish and be damned" '. But he would, I think, have made a distinction between suppressing truth and pursuing new truths at any human cost. He would, I suspect, have said that if man's endurance as a more or less decent creature depends on leaving certain doors closed, then so be it. He was, we remember, a socialist.

But the other answer is also possible. We hear it, unquenchable, from the dark of Thales' well, and from the blood-stained garden in Syracuse. It says to us that truth matters more than man. That it is more interesting than he, even when, perhaps especially when, it puts in question his own survival. I believe that truth has a future. Whether we do is less certain. But man alone can suppose this. And it is this supposition, first put forward in the Mediterranean world some three thousand years ago, which is the mark of his glory.

Marie Stopes

An extract from *Contraception*, 1932

The extraordinary parallel between the language and kind of argument used by those who objected to chloroform with that used by those who to-day oppose contraception on 'religious' grounds is so remarkable that there is little doubt that in another twenty years or less those same 'arguments' will be used and those same objurgations hurled at some other advance of scientific alleviation of human suffering, and that no priest or cleric will dare to inveigh against birth control then, just as to-day none dares to repeat the sermons of his predecessors against chloroform.

A. Strange

Extracts from 'On the necessity for a permanent commission on state scientific questions', 1871

The duty of the government with respect to science is one of the questions of the day. No question of equal importance has, perhaps, been more carelessly considered and more heedlessly postponed than this. And, now that a hearing has been obtained for it, neither the governing class nor the masses are qualified to discuss it intelligently; the governing class, because it is for the most part composed of men in whose education, as even the highest education was conducted 30 to 50 years ago, science occupied an insignificant place; and the masses, because they may be taken to be virtually destitute of scientific knowledge. Those who wield, and those who confer, the powers of government being alike incapable of dealing with this question, it devolves on another section of the community to urge its claims to attention. . . .

I must guard myself against the supposition that the proposal I have here advocated comprises all that is necessary for the efficient administration of scientific State affairs. It is only one part of a great system that has to be created. Other parts of the system will, no doubt, receive due attention from the Royal Commission now considering them. But there is one part so important that I feel called on to name it – I mean the appointment of a Minister of Science. He need not necessarily be exclusively devoted to science; he might, perhaps, with advantage, have charge of education and fine arts also; but some one in Parliament, directly representing the scientific branches of the national services has become absolutely indispensable. Another urgent want which, as its scientific character is not purely physical, will probably not be dealt with by the Royal Commission on Science, is that of a high war council – a council of naval and military officers of the greatest professional attainments and distinction, constituted for the purpose of advising the government on the highest problems of strategical science. At present we have not a vestige of anything of the kind, and are consequently, as a military nation, almost destitute of the basis of the military art.

When we have all scientific national institutions under one Minister of State, advised by a permanent, independent, and highly-qualified consultative body; when we have a similar body to advise the Ministers of War and Marine in strategical science, then the fact that, in accordance with our marvellous constitution, these ministers must almost necessarily be men without pretension to a knowledge of the affairs which they administer need cause us no alarm. When these combinations have been, as they assuredly will be, sooner or later, effected, the wealth, resources, and intelligence of the nation, having due scope, will render us unapproachable in the arts of peace and unconquerable in war, but not till then.

Lewis Thomas

'Notes of a biology watcher: to err is human', from *New England Journal of Medicine*, 1974

Statistically, the probability of any one of us being here is so small that you'd think the mere fact of existing would keep us all in a contented dazzlement of surprise. We are alive against the stupendous odds of genetics, infinitely outnumbered by all the alternates who might, except for luck, be in our places.

Even more astounding is our statistical improbability in physical terms. The normal, predictable state of matter throughout the universe is randomness, a relaxed sort of equilibrium, with atoms and their particles scattered around in an amorphous muddle. We, in brilliant contrast, are completely organized structures, squirming with information at every covalent bond. We make our living by catching electrons at the moment of their excitement by solar photons, swiping the energy released at the instant of each jump and storing it up in intricate loops for ourselves. We violate probability, by our nature. To be able to do this systemically, and in such wild varieties of form, from viruses to whales, is extremely unlikely; to have sustained the effort successfully for the several billion years of our existence, without drifting back into randomness, was nearly a mathematical impossibility.

Add to this the biological improbability that makes each member of our own species unique. Everyone is one in 3 billion at the moment, which describes the odds. Each of us is a self-contained, free-standing individual, labeled by specific protein configurations at the surfaces of cells, identifiable by whorls of fingertip skin, maybe even by special medleys of fragrance. You'd think we'd never stop dancing.

Perhaps it is not surprising that we do not live more surprised. After all, we are used to unlikelihood. Being born into it, raised in it, we become acclimated to the altitude, like natives in the Andes. Moreover, we all know that the astonishment is transient, and sooner or later our particles will all go back to being random.

Also, there are reasons to suspect that we are really not the absolute, pure entities that we seem. We have some sense of ordinariness, and it tends to diminish our surprise. Despite all the evidences of biological privacy in our cells and tissues (to the extent that a fragment of cell membrane will be recognized and rejected between any conceivable pairs among the 3 billion, excepting identical twins), there is a certain slippage in our brains. No one, in fact, can lay claim with certainty to his own mind with anything like the specificity stipulated by fingerprints or tissue antigens.

The human brain is the most public organ on the face of the earth, open to everything, sending out messages to everything. To be sure, it is hidden away in bone and conducts internal affairs in secrecy, but virtually all the business is the direct result of thinking that has already occurred in other minds. We pass thoughts around, from mind to mind, so compulsively and with such speed that the brains of mankind often appear, functionally, to be undergoing fusion.

This is, when you think about it, really amazing. The whole dear notion of one's own Self – marvelous old free-willed, free-enterprising, autonomous, independent, isolated island of a Self – is a myth.

We do not yet have a science strong enough to displace the myth. If you could label, by some equivalent of radioactive isotopes, all the bits of human thought that are constantly adrift, like plankton, all around us, it might be possible to discern some sort of systematic order in the process, but, as it is, it seems almost entirely random. There has to be something wrong with this view. It is hard to see how we could be in possession of an organ so complex and intricate and, as it occasionally reveals itself, so powerful, and be using it on such a scale just for the production of a kind of background noise. Somewhere, obscured by the snatches of conversation, pages of old letters, bits of books and magazines, memories of old movies, and the disorder of radio and television, there ought to be more intelligible signals.

Or perhaps we are only at the beginning of learning to use the system, with almost all our evolution as a species still ahead of us. Maybe the thoughts we generate today and flick around from mind to mind, like the jokes that turn up simultaneously at dinner parties in Hong Kong and Boston, or the sudden changes in the way we wear our hair, or all the popular love songs, are the primitive precursors of more complicated, polymerized structures that will come later, analogous to the prokaryotic cells that drifted through shallow pools in the early days of biological evolution. Later, when the time is right, there may be fusion and symbiosis among the bits, and then we will see eukaryotic thought, metazoans of thought, huge interliving coral shoals of thought.

The mechanism is there, and there is no doubt that it is already capable of functioning, even though the total yield thus far seems to consist largely of bits. After all, it has to be said that we've been at it for only the briefest time in evolutionary terms, a few thousand years out of billions, and during most of this time the scattered aggregates of human thought have been located patchily around the earth. There may be some laws about this kind of communication, mandating a critical density and mass before it can function with efficiency. Only in this century have we been brought close enough to each other, in great numbers, to begin the fusion around the earth, and from now on the process may move very rapidly.

There is, if it goes well, quite a lot to look forward to. Already, by luck, we have seen the assembly of particles of exchanged thought [grow] into today's structures of art and science. It is done by simply passing the bits around from mind to mind, until something like natural selection makes the final selection, all on grounds of fitness.

The real surprises, which set us back on our heels when they occur, will always be the mutants. We have already had a few of these, sweeping across the field of human thought periodically, like comets. They have slightly different receptors for the information cascading in from other minds, and slightly different machinery for processing it, so that what comes out to rejoin the flow is novel, and filled with new sorts of meaning. Bach was able to do this, and what emerged in the current were primordia in music. In this sense, the Art of Fugue and the St Matthew Passion were, for the evolving organism of human thought, feathered wings, apposing thumbs, new layers of frontal cortex.

But we may not be so dependent on mutants from here on, or perhaps there are more of them around than we recognize. What we need is more crowding, more unrestrained and obsessive communication, more open channels, even more noise, and a bit more luck. We are simultaneously participants and bystanders, which is a puzzling role to play. As participants, we have no choice in the matter; this is what we do as a species. As bystanders, stand back and give it room is my advice.

'Biomedical science and human health: the long-range prospect', from *Daedalus*, 1977

It is customary to place the date for the beginnings of modern medicine somewhere in the mid-1930s, with the entry of the sulfonamides and penicillin into the pharmacopoeia, and it is usual to ascribe to these events the force of a revolution in medical practice. This is what things seemed like at the time. Medicine was upheaved, revolutionized indeed. Therapy had been discovered for great numbers of patients whose illnesses had previously been untreatable. Cures were now available. As we saw it then, it seemed a totally new world. Doctors could now *cure* diseases, and this was astonishing, most of all to the doctors themselves.

It was, no doubt about it, a major occurrence in medicine, and a triumph for biological science applied to medicine – but perhaps not a revolution after all, looking back from this distance. For the real revolution in medicine, which set the stage for antibiotics and whatever else we have in the way of effective therapy today, had already occurred 100 years before penicillin. It did not begin with the introduction of science into medicine. That came years later. Like a good many revolutions, this one began with the destruction of dogma. It was discovered, sometime in the 1830s, that the greater part of medicine was nonsense.

The history of medicine has never been a particularly attractive subject in medical education, and one reason for this is that it is so unrelievedly deplorable a story. For century after century, all the way into the remote millennia of its origins, medicine got along by sheer guesswork and the crudest sort of empiricism. It is hard to conceive of a less scientific enterprise among human endeavors. Virtually anything that could be thought up for the treatment of disease was tried out at one time or another, and, once tried, lasted decades or even centuries before being given up. It was, in retrospect, the most frivolous and irresponsible kind of human experimentation, based on nothing but trial and error and usually resulting in precisely that sequence. Bleeding, purging, cupping, the administration of infusions of every known plant, solutions of every known metal, every conceivable diet including total fasting, most of these based on the weirdest imaginings about the cause of disease, concocted out of nothing but thin air – this was the heritage of medicine up until a little over a century ago. It is astounding that the profession survived so long, and got away with so much with so little outcry. Almost everyone seems to have been taken in. Evidently one had to be a born skeptic, like Montaigne, to see through the old nonsense; but even Montaigne, who wrote scathingly about the illnesses caused by doctoring centuries before Ivan Illich, had little effect. Most people were convinced of the magical powers of medicine and put up with it.

Then, sometime in the early nineteenth century, it was realized by a few of the leading figures in medicine that almost all of the complicated treatments then available for disease did not really work, and the suggestion was made by several courageous physicians, here and abroad, that most of them actually did more harm than good. Simultaneously, the surprising discovery was made that certain diseases were self-limited, got better by themselves, possessed, so to speak, a 'natural history'. It is hard for us now to imagine the magnitude of this discovery and its effect on the practice of medicine. The long habit of medicine, extending back into the distant past, had been to treat everything with something, and it was taken for granted that every disease demanded treatment and might in fact end fatally if not treated. In a sober essay written on this topic in 1876, Professor Edward H. Clarke of Harvard reviewed what he regarded as the major scientific accomplishment of medicine in the preceeding fifty years, which consisted of studies proving that patients with typhoid and typhus fever could recover all by themselves, without medical intervention, and often did better for being untreated than when they received the bizarre herbs, heavy metals, and fomentations that were popular at that time. Delirium tremens, a disorder long believed to be fatal in all cases unless subjected to constant and aggressive medical intervention, was observed to subside by itself more readily in patients left untreated, with a substantialy improved rate of survival.

Gradually, over the succeeding decades, the traditional therapeutic ritual of medicine was given up, and what came to be called the 'art of medicine' emerged to take its place. In retrospect, this art was really the beginning of the science of medicine. It was based on meticulous, objective, even cool observations of sick people. From this endeavor we learned the details of the natural history of illness, so that, for example, it came to be understood that typhoid and typhus were really two entirely separate, unrelated disorders, with quite different causes. Accurate diagnosis became the central purpose and justification for medicine, and as the methods for diagnosis improved, accurate prognosis also became possible, so that patients and their families could be told not only the name of the illness but also, with some reliability, how it was most likely to turn out. By the time this century had begun, these were becoming generally accepted as the principal responsibilities of the physician. In addition, a new kind of much less ambitious and flamboyant therapy began to emerge, termed *supportive treatment*, and consisting in large part of plain common sense: good nursing care, appropriate bed rest, a sensible diet, avoidance of traditional nostrums and patent medicine, and a measured degree of trust that nature, in taking its course, would very often bring things to a satisfactory conclusion.

The doctor became a considerably more useful and respected professional. For all his limitations, and despite his inability to do much in the way of preventing or terminating illness, he could be depended on to explain things, to relieve anxieties, and to be on hand. He was trusted as an advisor and guide in difficult times, including the time of dying.

Meanwhile, starting in the last decade of the nineteenth century, the basic science needed for a future science of medicine got under way. The role of bacteria and viruses in illness was discerned, and research on the details of this connection began in earnest. The major pathogenic organisms, most notably the tubercle bacillus and the syphilis spirochete, were recognized for what they were and did. By the late 1930s this research had already paid off; the techniques of active and passive immunization had been worked out for diphtheria, tetanus, lobar pneumonia, and a few other bacterial infections; the taxonomy of infectious disease had become an orderly discipline; and the time was ready for sulfanilamide, penicillin, streptomycin, and all the rest. But it needs emphasizing that it took about fifty years of concentrated effort in basic research to reach this level; if this research had not been done we could not have guessed that streptococci and pneumococci exist, and the search for antibiotics would have made no sense at all. Without the long, painstaking research on the tubercle bacillus, we would still be thinking that tuberculosis was due to night air, and we would still be trying to cure it by sunlight.

At that time, after almost a century of modified skepticism about therapy amounting finally to near nihilism, we abruptly entered a new era in which, almost overnight, it became possible with antibiotics to cure outright some of the most common and lethal illnesses of human beings – lobar pneumonia, meningitis, typhoid, typhus, tuberculosis, septicemias of various types. Only the virus diseases lay beyond reach, and even some of these were shortly to come under control – as in poliomyelitis and measles – by new techniques for making vaccines.

These events were simply overwhelming when they occurred. I was a medical student at the time of sulfanilamide and penicillin, and I remember the earliest reaction of flat disbelief concerning such things. We had given up on therapy, a century earlier. With a few exceptions which we regarded as anomalies, such as vitamin B for pellagra, liver extract for pernicious anemia, and insulin for diabetes, we were educated to be skeptical about the treatment of disease. Military tuberculosis and subacute bacterial endocarditis were fatal in 100 percent of cases, and we were convinced that the course of master diseases like these could never be changed, not in our lifetime or in any other.

Overnight, we became optimists, enthusiasts. The realization that disease could be turned around by treatment, provided that one knew enough about the underlying mechanism, was a totally new idea just forty years ago.

Most people have forgotten about that time, or are too young to remember it, and tend now to take such things for granted. They were born knowing about antibiotics, or the drugs simply fell by luck into their laps. We need reminding, now more than ever, that the capacity of medicine to deal with infectious disease was not a lucky fluke, nor was it something that happened simply as the result of the passage of time. It was the direct outcome of many years of hard work, done by imaginative and skilled scientists, none of whom had the faintest idea that penicillin and streptomycin lay somewhere in the decades ahead. It was basic science of a very high order, storing up a great mass of interesting knowledge for its own sake, creating, so to speak, a bank of information, ready for drawing on when the time for intelligent use arrived.

For example, it took a great deal of time, and work, before it could be understood that there were such things as hemolytic streptococci, that there were more than forty different serological types of the principal streptococcal species responsible for human disease, and that some of these were responsible for rheumatic fever and valvular heart disease. The bacteriology and immunology had to be done first, over decades, and by the early 1930s the work had progressed just far enough so that the connection between streptococcal infection and rheumatic fever could be perceived.

Not until this information was at hand did it become a certainty that rheumatic fever could be prevented, and with it a large amount of the chief heart disease affecting young people, if only a way could be found to prevent streptococcal infection. Similarly, the identification of the role of pneumococci in lobar pneumonia, of brucellae in undulant fever, typhoid bacilli in typhoid fever, the meningococcus in epidemic meningitis, required the sorting out and analysis of what seemed at the time an immensely complicated body of information. Most of the labor in infectious disease laboratories went into work of this kind in the first third of this century. When it was finished, the scene was ready for antibiotics.

What was not realized then – and is not fully realized even now – was how difficult it would be to accomplish the same end for the other diseases of man. We still have heart disease, cancer, stroke, schizophrenia, arthritis, kidney failure, cirrhosis, and the degenerative diseases associated with aging. All told there is a list of around twenty-five major afflictions of man in this country, and a still more formidable list of parasitic, viral, and nutritional diseases in the less developed countries of the world, which make up the unfinished agenda of modern biomedical science.

How does one make plans for science policy with such a list? The quick and easy way is to conclude that these diseases, not yet mastered, are simply beyond our grasp. The thing to do is to settle down with today's versions of science and technology, and make sure that our health care system is equipped to do the best it can in an imperfect world. The trouble with this approach is that we cannot afford it. The costs are already too high, and they escalate higher each year. Moreover, the measures available are simply not good enough. We cannot go on indefinitely trying to cope with heart disease by open-heart surgery, carried out at formidable expense after the disease has run its destructive course. Nor can we postpone such issues by oversimplifying the problems, which is what we do, in my opinion, by attributing so much of today's chronic and disabling disease to the environment, or to wrong ways of living. The plain fact of the matter is that we do not know enough about the facts of the matter, and we should be more open about our ignorance.

At the same time – and this will have a paradoxical sound – there has never been a period in medicine when the future has looked so bright. There is within medicine, somewhere beneath the pessimism and discouragement resulting from the disarray of the health care system and its stupendous cost, an undercurrent of almost outrageous optimism about what may lie ahead for the treatment of human disease if we can only keep learning. The scientists who do research on the cardiovascular system are entirely confident that they will soon be working close to the center of things, and they no longer regard the mechanisms of heart disease as impenetrable mysteries. The cancer scientists, for all their public disagreements about how best to organize their research, are in possession of insights into the intimate functioning of normal and neoplastic cells that were unimaginable a few years back. The eukaryotic cell – the cell with a true nucleus – has itself become a laboratory instrument almost as neat and handy as the bacterial cell became in the early 1950s, ready now to be used for elucidating the mechanisms by which genes are switched on or off as developing cells differentiate or, as in the case of cancer cells, de-differentiate. The ways in which carcinogenic substances, or viruses, or other factors still unrecognized intervene in the regulation of cell behavior represent problems still unsolved, but the problems themselves now appear to be approachable; with what has been learned in the past decade, they can now be worked on.

The neurobiologists can do all sorts of things in their investigation, and the brain is an organ different from what it seemed twenty-five years ago. Far from being an intricate but ultimately simplifiable mass of electronic circuitry governed by wiring diagrams, it now has the aspect of a fundamentally endocrine tissue, in which the essential reactions, the internal traffic of nerve impulses, are determined by biochemical activators and their suppressors. The technologies available for quantitative study of individual nerve cells are powerful and precise, and the work is now turning toward the functioning of collections of cells, the centers for visual and auditory perception and the like, because work at this level can now be done. It is difficult to think of problems that cannot be studied, now or ever. The matter of consciousness is argued over, naturally, as a candidate for perpetual unapproachability, but this has more the sound of a philosophical discussion. Nobody has the feeling any longer, as we used to believe, that we can never find out how the brain works.

The immunologists, the molecular biochemists, and the new generation of investigators obsessed with the structure and function of cell membranes have all discovered that they are really working together, along with the geneticists, on a common set of problems: how do cells and tissues become labeled for what they are, what are the forces that govern the orderly development and differentiation of tissues and organs, and how are errors in the process controlled?

There has never been a time like it, and I find it difficult to imagine that this tremendous surge of new information will terminate with nothing more than an understanding of how normal cells and tissues, and organisms, function. I regard it as a certainty that there will be uncovered, at the same time, detailed information concerning the mechanisms of disease.

The record of the past half-century has established, I think, two general principles about human disease. First, it is necessary to know a great deal about underlying mechanisms before one can really act effectively; one had to know that the pneumococcus causes lobar pneumonia before one could begin thinking about antibiotics. One did not have to know all the details, not even how the pneumococcus does its damage to the lungs; but one had to know that it was there, and in charge.

Second, for every disease there is a single key mechanism that dominates all others. If one can find it, and then think one's way around it, one can control the disorder. This generalization is harder to prove, and arguable – it is more like a strong hunch than a scientific assertion – but I believe that the record thus far tends to support it. The most complicated, multicell, multitissue, and multi-organ diseases I know of are tertiary syphilis, chronic tuberculosis, and pernicious anemia. In each, there are at least five major organs and tissues involved, and each appears to be affected by a variety of environmental influences. Before they came under scientific appraisal each was thought to be what we now call a 'multifactorial' disease, far too complex to allow for any single causative mechanism. And yet, when all the necessary facts were in, it was clear that by simply switching off one thing – the spirochete, the tubercle bacillus, or a single vitamin deficiency – the whole array of disordered and seemingly unrelated pathologic mechanisms could be switched off, at once.

I believe that a prospect something like this is the likelihood for the future of medicine. I have no doubt that there will turn out to be dozens of separate influences that can launch cancer, including all sorts of environmental carcinogens and very likely many sorts of virus, but I think there will turn out to be a single switch at the center of things, there for the finding. I think that schizophrenia will turn out to be a neurochemical disorder, with some central, single chemical event gone wrong. I think there is a single causative agent responsible for rheumatoid arthritis, which has not yet been found. I think that the central vascular abnormalities that launch coronary occlusion and stroke have not yet been glimpsed, but they are there, waiting to be switched on or off.

In short, I believe that the major diseases of human beings have become approachable biological puzzles, ultimately solvable. It follows from this that it is now possible to begin thinking about a human society relatively free of disease. This would surely have been an unthinkable notion a half century ago, and oddly enough it has a rather apocalyptic sound today. What will we do about dying, and about all that population, if such things were to come about? What can we die of, if not disease?

My response is that it would not make all that much difference. We would still age away and wear out, on about the same schedule as today, with the terminal event being more like the sudden disintegration and collapse all at once of Oliver Wendell Holmes' well-known one-hoss shay. The main effect – almost pure benefit, it seems to me – would be that we would not be beset and riddled by disease in the last decades of life, as most of us are today. We could become a health species, not all that different from the healthy stocks of domestic plants and animals that we already take for granted. Strokes, and senile dementia, and cancer, and arthritis are not natural aspects of the human condition, and we ought to rid ourselves of such impediments as quickly as we can.

There is another argument against this view of the future which needs comment. It is said that we are fundamentally fallible as organisms, prone to failure, and if we succeed in getting rid of one set of ailments there will always be other new diseases, now waiting out in the forest, ready to take their places. I do not know why this is said, for I can see no evidence that such a thing has ever happened. To be sure, we have a higher incidence of chronic illness among older people than we had in the early years of this century, but that is because more of us have survived to become older people. No new disease, so far as I know, has come in to take the place of diphtheria, or smallpox, or whooping cough, or poliomyelitis. Nature being inventive, we will probably always have the odd new illness turning up, but not in order to fill out some ordained, predestined quota of human maladies.

Indeed, the official public health tables of morbidity and mortality seem to be telling us this sort of thing already, even though, in all our anxiety, we seem unwilling to accept the news. We have already become in the Western world, on the record, the healthiest society in the history of humankind. Compared with a century ago, when every family was obliged to count on losing members throughout the early years of life, we are in a new world. A death in a young family has become a rare and dreadful catastrophe, no longer a commonplace event. Our estimated life expectancy, collectively, is longer this year than ever before in history. Part of this general and gradual improvement in health and survival is thanks to sanitary engineering, better housing, and, probably, more affluence, but a substantial part is also attributable, in recent years, to biomedical science. We have not done badly at all, and having begun so well I see no reason why we should not do even better in the future.

My argument about how to do this will come as no surprise. I say that we must continue doing biomedical research, on about the same scale and scope as in the past twenty years, with expansion and growth of the enterprise being dependent on where new leads seem to be taking us. It is an expensive undertaking, but still it is less than 3 percent of the total annual cost of today's health industry, which at last count was over $120 billion, and it is nothing like as expensive as trying to live with the halfway technologies we are obliged to depend on in medicine today; if we try to stay with these for the rest of the century the costs will go through the ionosphere.

But I should like to insert a qualification in this argument, which may be somewhat more of a surprise, coming from a doctor. I believe that the major research effort, and far and away the greatest investment for the future, must be in the broad area of basic biological science. Here and there, to be sure, there will be opportunities for productive applied science, comparable, say, to the making of polio vaccine or the devising of multidrug therapy for childhood leukemia, but these opportunities will not come often, nor can they be forced into existence before their time. The great need now, for the medicine of the future, is for more information at the most fundamental levels of the living process. We are nowhere near ready for large-scale programs of applied science in medicine, for we do not yet know enough.

Good applied science in medicine, as in physics, requires a high degree of certainty about the basic facts at hand, and especially about their meaning, and we have not yet reached this point for most of medicine. Nor can we predict at this stage, with much confidence, which particular items of new information, from which fields, are the likeliest to be relevant to particular disease problems. In this circumstance there has to be a certain amount of guessing, even gambling; and my own view is that the highest yield for the future will come from whatever fields are generating the most interesting, exciting, and surprising sorts of information – most of all, surprising.

It seems to me that the safest and most prudent of bets to lay money on is surprise. There is a very high probability that whatever astonishes us in biology today will turn out to be usable, and useful, tomorrow. This, I think, is the established record of science itself, over the past two hundred years, and we ought to have more confidence in the process. It worked this way for the beginnings of chemistry; we obtained electricity in this manner; using surprise as a guide we progressed from Newtonian physics to electromagnetism, to quantum mechanics and contemporary geophysics and cosmology. In biology, evolution and genetics were the earliest big astonishments, but what has been going on in the past quarter century is simply flabbergasting. For medicine, the greatest surprises lie still ahead of us, but they are there, waiting to be discovered or stumbled over, sooner or later.

I am arguing this way from the most practical, down-to-earth, pragmatic point of view. This kind of science is most likely, in the real world, to lead to significant improvements in human health, and at low cost. This is a point worth further emphasis, by the way: When medicine has really succeeded brilliantly in technology, as in immunization, for example, or antibiotics, or nutrition, or endocrine replacement therapy, so that the therapeutic measures can be directed straight at the underlying disease mechanism and are decisively effective, the cost is likely to be very low indeed. It is when our technologies have to be applied halfway along against the progress of disease, or must be brought in after the fact to shore up the loss of destroyed tissue, that health care becomes enormously expensive. The deeper our understanding of a disease mechanism, the greater are our chances of devising direct and decisive measures to prevent disease, or to turn it around before it is too late.

So much for the practical side of the argument. We need much more basic science for the future of human health, and I will leave the matter there.

But I have one last thing to say about biological science. Even if I should be wrong about some of these predictions, and it turns out that we can blunder our way into treating or preventing one disease or another without understanding the process (which I will not believe until it happens), and if we continue to invest in biological science anyway, we cannot lose. The Congress, in its wisdom, can't lose. The public can't lose.

Here is what I have in mind.

These ought to be the best times for the human mind, but is not so. All sorts of things seem to be turning out wrong, and the century seems to be slipping through our fingers here at the end, with almost all promises unfilled. I cannot begin to guess at all the causes of our cultural sadness, not even the most important ones, but I can think of one thing that is wrong with us and eats away at us: we do not know enough about ourselves. We are ignorant about how we work, about where we fit in, and most of all about the enormous, imponderable system of life in which we are embedded as working parts. We do not really understand nature, at all. Not to downgrade us; we have come a long way indeed, just to have learned enough to become conscious of our ignorance. It is not so bad a thing to be totally ignorant; the hard thing is to be part way along toward real knowledge, far enough to be aware of being ignorant. It is embarrassing and depressing, and it is one of our troubles today.

It is a new experience for all of us. Only two centuries ago we could explain everything about everything, out of pure reason, and now most of that elaborate and harmonious structure has come apart before our eyes. We are *dumb*.

This is, in a certain sense, a health problem after all. For as long as we are bewildered by the mystery of ourselves, and confused by the strangeness of our uncomfortable connection to all the rest of life, and dumbfounded by the inscrutability of our own minds, we cannot be said to be healthy animals in today's world.

We need to know more. To come to realize this is what this seemingly inconclusive century has been all about. We have discovered how to ask important questions, and now we really do need, as an urgent matter, for the sake of our civilization, to obtain some answers. We now know that we cannot do this any longer by searching our minds, for there is not enough there to search, nor can we find the truth by guessing at it or by making up stories for ourselves. We cannot stop where we are, stuck with today's level of understanding, nor can we go back. I do not see that we have a real choice in this, for I can see only the one way ahead. We need science, more and better science, not for its technology, not for leisure, not even for health or longevity, but for the hope of wisdom which our kind of culture must acquire for its survival.

D'Arcy Wentworth Thompson

Part of the opening chapter of *On Growth and Form*, 1942

As soon as we adventure on the paths of the physicist, we learn to *weigh* and to *measure*, to deal with time and space and mass and their related concepts, and to find more and more our knowledge expressed and our needs satisfied through the concept of *number*, as in the dreams and visions of Plato and Pythagoras; for modern chemistry would have gladdened the hearts of those great philosophic dreamers. Dreams apart, numerical precision is the very soul of science, and its attainment affords the best, perhaps the only criterion of the truth of theories and the correctness of experiments. . . . So said Sir John Herschel, a hundred years ago; and Kant had said that it was Nature herself, and not the mathematician, who brings mathematics into natural philosophy.

But the zoologist or morphologist has been slow, where the physiologist has long been eager, to invoke the aid of the physical or mathematical sciences; and the reasons for this difference lie deep, and are partly rooted in old tradition and partly in the diverse minds and temperaments of men. To treat the living body as a mechanism was repugnant, and seemed even ludicrous, to Pascal . . .; and Goethe, lover of nature as he was, ruled mathematics out of place in natural history. Even now the zoologist has scarce begun to dream of defining in mathematical language even the simplest organic forms. When he meets with a simple geometrical construction, for instance in the honeycomb, he would fain refer it to psychical instinct, or to skill and ingenuity, rather than to the operation of physical forces or mathematical laws; when he sees in snail, or nautilus, or tiny foraminiferal or radiolarian shell a close approach to sphere or spiral, he is prone of old habit to believe that after all it is something more than a spiral or a sphere, and that in this 'something more' there lies what neither mathematics nor physics can explain. In short, he is deeply reluctant to compare the living with the dead, or to explain by geometry or by mechanics the things which have their part in the mystery of life. Moreover he is little inclined to feel the need of such explanations, or of such extension of his field of thought. He is not without some justification if he feels that in admiration of nature's handiwork he has an horizon open before his eyes as wide as any man requires. He has the help of many fascinating theories within the bounds of his own science, which, though a little lacking in precision, serve the purpose of ordering his thoughts and of suggesting new objects of enquiry. His art of classification becomes an endless search after the blood-relationships of things living and the pedigrees of things dead and gone. The facts of embryology record for him (as Wolff, von Baer and Fritz Müller proclaimed) not only the life-history of the individual but the ancient annals of its race. The facts of geographical distribution or even of the migration of birds lead on and on to speculations regarding lost continents, sunken islands, or bridges across ancient seas. Every nesting bird, every ant-hill or spider's web, displays its psychological problems of instinct or intelligence. Above all, in things both great and small, the naturalist is rightfully impressed and finally engrossed by the peculiar beauty which is manifested in apparent fitness or 'adaptation' – the flower for the bee, the berry for the bird.

Some lofty concepts, like space and number, involve truths remote from the category of causation; and here we must be content, as Aristotle says, if the mere facts be known. . . . But natural history deals with ephemeral and accidental, not eternal nor universal things; their causes and effects thrust themselves on our curiosity, and become the ultimate relations to which our contemplation extends. . . .

Time out of mind it has been by way of the 'final cause', by the teleological concept of end, of purpose or of 'design', in one of its many forms (for its moods are many), that men have been chiefly wont to explain the phenomena of the living world; and it will be so while men have eyes to see and ears to hear withal. With Galen, as with Aristotle, . . . it was the physician's way; with John Ray, . . . as with Aristotle, it was the naturalist's way; with Kant, as with Aristotle, it was the philosopher's way. It was the old Hebrew way, and has its splendid setting in the story that God made 'every plant of the field before it was in the earth, and every herb of the field before it grew'. It is a common way, and a great way; for it brings with it a glimpse of a great vision, and it lies deep as the love of nature in the hearts of men.

The argument of the final cause is conspicuous in eighteenth-century physics, half overshadowing the 'efficient' or physical cause in the hands of such men as Euler, . . . or Fermat or Maupertuis, to whom Leibniz . . . had passed it on. Half overshadowed by the mechanical concept, it runs through Claude Bernard's *Leçons sur les phénomènes de la Vie*, . . . and abides in much of modern physiology. . . . Inherited from Hegel, it dominated Oken's *Naturphilosophie* and lingered among his later disciples, who were wont to liken the course of organic evolution not to the straggling branches of a tree, but to the building of a temple, divinely planned, and the crowning of it with its polished minarets. . . .

It is retained, somewhat crudely, in modern embryology, by those who see in the early processes of growth a significance 'rather prospective than retrospective', such that the embryonic phenomena must 'be referred directly to their usefulness in building up the body of the future animal . . .': – which is no more, and no less, than to say, with Aristotle, that the organism is the τέλος, or final cause, of its own processes of generation and development. It is writ large in that Entelechy . . . which Driesch rediscovered, and which he made known to many who had neither learned of it from Aristotle, nor studied it with Leibniz, nor laughed at it with Rabelais and Voltaire. And, though it is in a very curious way, we are told that teleology was 'refounded, reformed and rehabilitated' by Darwin's concept of the origin of species, . . . for, just as the older naturalists held (as Addison . . . puts it) that 'the make of every kind of animal is different from that of every other kind; and yet there is not the least turn in the muscles, or twist in the fibres of any one, which does not render them more proper for that particular animal's way of life than any other cut or texture of them would have been': so, by the theory of natural selection, 'every variety of form and colour was urgently and absolutely called upon to produce its title to existence either as an active useful agent, or as a survival' of such active usefulness in the past. But in this last, and very important case, we have reached a teleology without a τέλος, as men like Butler and Janet have been prompt to shew, an 'adaptation' without 'design', a teleology in which the final cause becomes little more, if anything, than the mere expression or resultant of a sifting out of the good from the bad, or of the better from the worse, in short of a process of mechanism. The apparent manifestations of purpose or adaptation become part of a mechanical philosophy, 'une forme méthodologique de connaissance,' . . . according to which 'la Nature agit toujours par les moyens les plus simples,' . . . and 'chaque chose finit toujours par s'accommoder à son milieu', as in the Epicurean creed or aphorism that Nature *finds a use* for everything. . . . In short, by a road which resembles but is not the same as Maupertuis's road, we find our way to the very world in which we are living, and find that, if it be not, it is ever tending to become, 'the best of all possible worlds'. . . .

But the use of the teleological principle is but one way, not the whole or the only way, by which we may seek to learn how things came to be, and to take their places in the harmonious complexity of the world. To seek not for ends but for antecedents is the way of the physicist, who finds 'causes' in what he has learned to recognise as fundamental properties, or inseparable concomitants, or unchanging laws, of matter and of energy. In Aristotle's parable, the house is there that men may live in it; but it is also there because the builders have laid one stone upon another. It is as a *mechanism*, or a mechanical construction, that the physicist looks upon the world; and

Democritus, first of physicists and one of the greatest of the Greeks, chose to refer all natural phenomena to mechanism and set the final cause aside.

Still, all the while, like warp and woof, mechanism and teleology are interwoven together, and we must not cleave to the one nor despise the other; for their union is rooted in the very nature of totality. We may grow shy or weary of looking to a final cause for an explanation of our phenomena; but after we have accounted for these on the plainest principles of mechanical causation it may be useful and appropriate to see how the final causes would tally with the other, and lead towards the same conclusion. . . . Maupertuis had little liking for the final cause, and shewed some sympathy with Descartes in his repugnance to its application to physical science. But he found at last, taking the final and the efficient causes one with another, that 'l'harmonie de ces deux attributs est si parfaite que sans doute tous les effets de la Nature se pourroient déduire de chacun pris séparément. Une Mécanique aveugle et nécessaire suit les dessins de l'Intelligence la plus éclairée et la plus libre. . . .' Boyle also, the Father of Chemistry, wrote, in his latter years, a *Disquisition about the Final Causes of Natural Things: Wherein it is Inquir'd Whether, And (if at all) With what Cautions, a Naturalist should admit Them?* He found 'that all consideration of final cause is not to be banished from Natural Philosophy . . .'; but on the other hand 'that the naturalist who would deserve that name must not let the search and knowledge of final causes make him neglect the industrious indagation of efficients. . . .' In our own day the philosopher neither minimises nor unduly magnifies the mechanical aspect of the Cosmos; nor need the naturalist either exaggerate or belittle the mechanical phenomena which are profoundly associated with Life, and inseparable from our understanding of Growth and Form.

Nevertheless, when philosophy bids us hearken and obey the lessons both of mechanical and of teleological interpretation, the precept is hard to follow: so that oftentimes it has come to pass, just as in Bacon's day, that a leaning to the side of the final cause 'hath intercepted the severe and diligent enquiry of all real and physical causes', and has brought it about that 'the search of the physical cause hath been neglected and passed in silence'. So long and so far as 'fortuitous variation . . .' and the 'survival of the fittest' remain engrained as fundamental and satisfactory hypotheses in the philosophy of biology, so long will these 'satisfactory and specious causes' tend to stay 'severe and diligent enquiry . . . to the great arrest and prejudice of future discovery'. Long before the great Lord Keeper wrote these words, Roger Bacon had shewn how easy it is, and how vain, to survey the operations of Nature and idly refer her wondrous works to chance or accident, or to the immediate interposition of God. . . .

The difficulties which surround the concept of ultimate or 'real' causation, in Bacon's or Newton's sense of the word, the insuperable difficulty of giving any just and tenable account of the relation of cause and effect from the empirical point of view, need scarcely hinder us in our physical enquiry. As students of mathematical and experimental physics we are content to deal with those antecedents, or concomitants, of our phenomena without which the phenomenon does not occur – with causes, in short, which, *aliae ex aliis aptae et necessitate nexae*, are no more, and no less, than conditions *sine qua non*. Our purpose is still adequately fulfilled: inasmuch as we are still enabled to correlate, and to equate, our particular phenomena with more and more of the physical phenomena around, and so to weave a web of connection and interdependence which shall serve our turn, though the metaphysician withhold from that interdependence the title of causality. . . . We come in touch with what the schoolmen called a *ratio cognoscendi*, though the true *ratio efficiendi* is still enwrapped in many mysteries. And so handled, the quest of physical causes merges with another great Aristotelian theme – the search for relations between things apparently disconnected, and for 'similitude in things to common view unlike. . . .' Newton did not shew the cause of the apple falling, but he shewed a similitude ('the more to increase our wonder, with an apple') between the apple and the stars. . . . By doing so he turned old facts into new knowledge; and was well content if he could bring diverse phenomena under 'two or three Principles of Motion' even 'though the Causes of these Principles were not yet discovered'.

Moreover the naturalist and the physicist will continue to speak of 'causes', just as of old, though it may be with some mental reservations: for, as a French philosopher said in a kindred difficulty: 'ce sont là des manières de s'exprimer, et si elles sont interdites il faut renoncer à parler de ces choses.'

The search for differences or fundamental contrasts between the phenomena of organic and inorganic, of animate and inanimate, things, has occupied many men's minds, while the search for community of principles or essential similitudes has been pursued by few; and the contrasts are apt to loom too large, great though they may be. M. Dunan, discussing the *Problème de la Vie*, . . . in an essay which M. Bergson greatly commends, declares that 'les lois physico-chimiques sont aveugles et brutales; là où elles règnent seules, au lieu d'un ordre et d'un concert, il ne peut y avoir qu'incohérence et chaos.' But the physicist proclaims aloud that the physical phenomena which meet us by the way have their forms not less beautiful and scarce less varied than those which move us to admiration among living things. The waves of the sea, the little ripples on the shore, the sweeping curve of the sandy bay between the headlands, the outline of the hills, the shape of the clouds, all these are so many riddles of form, so many problems of morphology, and all of them the physicist can more or less easily read and adequately solve: solving them by reference to their antecedent phenomena, in the material system of mechanical forces to which they belong, and to which we interpret them as being due. They have also, doubtless, their *immanent* teleological significance; but it is on another plane of thought from the physicist's that we contemplate their intrinsic harmony . . . and perfection, and 'see that they are good'.

Nor is it otherwise with the material forms of living things. Cell and tissue, shell and bone, leaf and flower, are so many portions of matter, and it is in obedience to the laws of physics that their particles have been moved, moulded and conformed. . . . They are no exception to the rule that Θεòς ἀεὶ γεωμετρεî. Their problems of form are in the first instance mathematical problems, their problems of growth are essentially physical problems, and the morphologist is, *ipso facto*, a student of physical science. He may learn from that comprehensive science, as the physiologists have not failed to do, the point of view from which her problems are approached, the quantitative methods by which they are attacked, and the wholesome restraints under which all her work is done. He may come to realise that there is no branch of mathematics, however abstract, which may not some day be applied to phenomena of the real world. . . . He may even find a certain analogy between the slow, reluctant expression of physical laws to vital phenomena and the slow triumphant demonstration by Tycho Brahé, Copernicus, Galileo and Newton (all in opposition to the Aristotelian cosmogony), that the heavens are formed of like substance with the earth, and that the movements of both are subject to the selfsame laws.

Organic evolution has its physical analogue in the universal law that the world tends, in all its parts and particles, to pass from certain less probable to certain more probable configurations or states. This is the second law of thermodynamics. It has been called *the law of evolution of the world* . . .; and we call it, after Clausius, the Principle of *Entropy*, which is a literal translation of *Evolution* into Greek.

The introduction of mathematical concepts into natural science has seemed to many men no mere stumbling-block, but a very parting of the ways. Bichat was a man of genius, who did immense service to philosophical anatomy, but, like Pascal, he utterly refused to bring physics or mathematics into biology: 'On calcule le retour d'un comète, les résistances d'un fluide parcourant un canal inerte, la vitesse d'un projectile, etc.; mais calculer avec Borelli la force d'un muscle, avec Keil la vitesse du sang, avec Jurine, Lavoisier et d'autres la quantité d'air entrant dans le poumon, c'est bâtir sur un sable mouvant un édifice solide par lui-même, mais qui tombe bientôt faute des base assurée. . . .' Comte went further still, and said that every attempt to introduce mathematics into chemistry must be deemed profoundly irrational, and contrary to the whole spirit of the science. . . . But the great makers of modern science have all gone the other way. Von Baer, using a bold metaphor, thought that it might become possible 'die bildenden Kräfte des thierischen Körpers auf die allgemeinen Kräfte oder *Lebenserscheinungen des Weltganzes* zurückzuführen. . . .' Thomas Young shewed, as Borelli had done, how physics may subserve anatomy; he learned from the heart and arteries that 'the mechanical motions which take place in an animal's body are regulated by the same general laws as the motions of inanimate bodies. . . .' And Theodore Schwann said plainly, a hundred years ago, 'Ich wiederhole übrigens dass, wenn hier von einer physikalischen Erklärung der organischen Erscheinungen die Rede ist, darunter nicht nothwendig eine Erklärung durch die bekannten physikalischen Kräfte . . . zu verstehen ist, sondern überhaupt eine Erklärung durch Kräfte, die nach strengen Gesetzen der blinden Nothwendigkeit wie die physikalischen Kräfte wirken, mögen diese Kräfte auch in der anorganischen Natur auftreten oder nicht. . . .'

Helmholtz, in a famous and influential lecture, and surely with these very words of Schwann's in mind, laid it down as the fundamental principle of physiology that 'there may be other agents acting in the living body than those agents which act in the inorganic world; but these forces, so far as they cause chemical and mechanical influence in the body, must be *quite of the same character* as inorganic forces: in this, at least, that their effects must be ruled by necessity, and must always be the same when acting under the same conditions; and so there cannot exist any arbitrary choice in the direction of their actions.' It follows further that, like the other 'physical' forces, they must be subject to mathematical analysis and deduction. . . .

So much for the physico-chemical problems of physiology. Apart from these, the road of physico-mathematical or dynamical investigation in morphology has found few to follow it; but the pathway is old. The way of the old Ionian physicians, of Anaxagoras . . ., of Empedocles and his disciples in the days before Aristotle, lay just by that highway side. It was Galileo's and Borelli's way; and Harvey's way, when he discovered the circulation of the blood. . . . It was little trodden for long afterwards, but once in a while Swammerdam and Réaumur passed thereby. And of later years Moseley and Meyer, Berthold, Errera and Roux have been among the little band of travellers. We need not wonder if the way be hard to follow, and if these wayfarers have yet gathered little. A harvest has been reaped by others, and the gleaning of the grapes is slow.

It behoves us always to remember that in physics it has taken great men to discover simple things. They are very great names indeed which we couple with the explanation of the path of a stone, the droop of a chain, the tints of a bubble, the shadows in a cup. It is but the slightest adumbration of a dynamical morphology that we can hope to have until the physicist and the mathematician shall have made these problems of ours their own, or till a new Boscovich shall have written for the naturalist the new *Theoria Philosophiae Naturalis*.

How far even then mathematics will suffice to describe, and physics to explain, the fabric of the body, no man can foresee. It may be that all the laws of energy, and all the properties of matter, and all the chemistry of all the colloids are as powerless to explain the body as they are impotent to comprehend the soul. For my part, I think it is not so. Of how it is that the soul informs the body, physical science teaches me nothing; and that living matter influences and is influenced by mind is a mystery without a clue. Consciousness is not explained to my comprehension by all the nerve-paths and neurones of the physiologist; nor do I ask of physics how goodness shines in one man's face, and evil betrays itself in another. But of the construction and growth and working of the body, as of all else that is of the earth earthy, physical science is, in my humble opinion, our only teacher and guide.

Often and often it happens that our physical knowledge is inadequate to explain the mechanical working of the organism; the phenomena are superlatively complex, the procedure is involved and entangled, and the investigation has occupied but a few short lives of men. When physical science falls short of explaining the order which reigns throughout these manifold phenomena – an order more characteristic in its totality than any of its phenomena in themselves – men hasten to invoke a guiding principle, an entelechy, or call it what you will. But all the while no physical law, any more than gravity itself, not even among the puzzles of stereo-chemistry or of physiological surface-action and osmosis, is known to be transgressed by the bodily mechanism.

Some physicists declare, as Maxwell did, that atoms or molecules more complicated by far than the chemist's hypotheses demand, are requisite to explain the phenomena of life. If what is implied be an explanation of psychical phenomena, let the point be granted at once; we may go yet further and decline, with Maxwell, to believe that anything of the nature of physical complexity, however exalted, could ever suffice. Other physicists, like Auerbach . . ., or Larmor . . ., or Joly . . ., assure us that our laws of thermodynamics do not suffice, or are inappropriate, to explain the maintenance, or (in Joly's phrase) the accelerative absorption, of the bodily energies, the retardation of entropy, and the long battle against the cold and darkness which is death. With these weighty problems I am not for the moment concerned. My sole purpose is to correlate with mathematical statement and physical law certain of the simpler outward phenomena of organic growth and structure or form, while all the while regarding the fabric of the organism, *ex hypothesi*, as a material and mechanical configuration. This is my purpose here. But I would not for the world be thought to believe that this is the only story which Life and her Children have to tell. One does not come by studying living things for a lifetime to suppose that physics and chemistry can account for them all. . . .

Physical science and philosophy stand side by side, and one upholds the other. Without something of the strength of physics philosophy would be weak; and without something of philosophy's wealth physical science would be poor. 'Rien ne retirera du tissu de la science les fils d'or que la main du philosophe y a introduits. . . .' But there are fields where each, for a while at least, must work alone; and where physical science reaches its limitations physical science itself must help us to discover. Meanwhile the appropriate and legitimate postulate of the physicist, in approaching the physical problems of the living body, is that with these physical phenomena no alien influence interferes. But the postulate, though it is certainly legitimate, and though it is the proper and necessary prelude to scientific enquiry, may some day be proven to be untrue; and its disproof will not be to the physicist's confusion, but will come as his reward. In dealing with forms which are so concomitant with life that they are seemingly controlled by life, it is in no spirit of arrogant assertiveness if the physicist begins his argument, after the fashion of a most illustrious exemplar, with the old formula of scholastic challenge: *An Vita sit? Dico quod non.* . . .

Meredith Thring

'Scientist Oath', from the *New Scientist*, 1971

'I vow to strive to apply my professional skills only to projects which, after conscientious examination, I believe to contribute to the goal of co-existence of all human beings in peace, human dignity and self-fulfilment.

I believe that this goal requires the provision of an adequate supply of the necessities of life (good food, air, water, clothing and housing, access to natural and man-made beauty), education, and opportunities to enable each person to work out for himself his life objectives and to develop creativeness and skill in the use of hands as well as head.

I vow to struggle through my work to minimise danger; noise; strain or invasion of privacy of the individual; pollution of earth, air or water; destruction of natural beauty, mineral resources and wildlife.'

Colin Tudge

'The great cuisines are waiting', chapter 7 of *The Famine Business*, 1977

Nutritionists are wont to paint a picture of the healthy life in which masochism stands well to the fore. School marms exhort us to eat up our overdone spinach, while vegans regale us with an endless round of muesli and nutburgers. Gastronomes too often present food as pure sensation, with little thought for the whole man and none for the world in which he lives. Thus has the cause of good nutrition been perverted by the faddist and the bossy dietician, and gastronomy shamefully corrupted by the food snob.

But the products of our rational agriculture are not the foundations of a mere food fad. We have grains, beans, and potatoes for basic nourishment; plenty of fresh vegetables and fruit, and modest but appreciable amounts of lean meat, from many species and all parts of the animal, to fill in nutritional cracks and add the essential element of pleasure. We have home-grown herbs – for many of which the harshest climates are adequate, and for others of which microclimates, found almost everywhere, will suffice. We have small quantities of imported spice and fruit – those high-value, small-volume crops that tropical countries can produce for monetary gain without perverting the structure of their own rational agriculture.

We have, indeed, the basis for all the world's great peasant cuisines: not surprising, since those cuisines have been evolving for at least 15,000 years to exploit the products of modest agriculture to the best advantage. And from peasant cuisines have grown the best of grande cuisine, as the great cooks – Fanny Craddock, Elizabeth David, Jane Grigson, Robert Carrier, Margaret Costa, Kenneth Lo – constantly emphasize: and we can say goodbye to much of what now passes as grande cuisine, overladen with double cream, without a tear. We have nothing to fear. We have only to re-learn how to cook.

In the great cuisines grain is often used as grain – as in the rice of South-East Asia, the pearled barley in British stews or as oatmeal in Scotland. Time and again we see the more or less unprocessed grain – used after no more than a cursory milling – garnished with small amounts of meat: precisely the meat–plant protein balance now recommended to make best use of the former and guarantee the nutritional quality of the latter. Hence the pulaos and biryanis of India; the dolmades – finely comminuted meat with spice and rice in wine leaves – of Greece; and its almost exact equivalent in Scotland, the haggis, where liver, lung and what you will are married with oatmeal. Haggis is both a gastronomic and nutritional triumph – and loses nothing, except in ethnic purity, in being matched with chips (French fries) in Glasgow. In general the cuisine of Scotland, developed by an ingenious people through hundreds of years of extreme austerity, deserves close study.

Of the major grains – rice, wheat and maize – only wheat seems to me to lack something when cooked in its pristine state. It tastes well, though powerful, but takes a long time to cook and tends to remain too chewy (for my taste). I can find little evidence that wheat, as wheat, has extensively featured in peasant cuisine. But agriculturalists – who in Britain are now encouraged by the government to make their research more relevant to people's needs – might do worse than develop strains that can be cooked as easily as rice.

However, I have found that if half the wheat grains are first smashed in the spice grinder, mixed with the rest of the whole 'berries', and then broiled in a dry skillet till it pops to bring out the nuttiness; then lightly fried and then boiled in about three-and-a-half times the volume of water for an hour or so, the result is a good complement to, for example, mackerel. Macrobiotic cooks make much of wheat: their recipes, involving spices and vegetables, are worth a look. The issue is not vital – but it is irritating that the world's great cuisines have apparently made so little of what could be an interesting resource.

Grain in a slightly more comminuted form, or deliberately bred for glutinousness, becomes the basis of the Chinese 'rice gruels', or congees, tricked out with liver, shellfish, cabbage or what you will; of the Chinese thick sweet-corn soups; and of Scottish porridge, made with oats. Again it would be nice if wheat could be used in this way, and the range of oats extended. The congee is a delightful form, and rational agriculturalists should not need recourse to imported rice.

Grind grain a little more – into flour – and we move into a whole new area. The list of cakes, pastries, noodles, and bread is far too extensive to treat in detail. But let us note that the Englishman's almost exclusive concentration on wheat flour is a sad restriction. It is less notable in the US, where corn and rye-bread flourish; or in Scotland, where the oatcake in its many variants lives on.

Note too how the products of the baker's shop have been sadly undermined by the protein myth. The conventional view of the meat pie, for example, is that the pastry acts merely as 'stodge', to keep the gravy in. The modern view is that meat and pastry protein complement each other, while the starch in the pastry provides the energy that 'spares' the protein. In a sweet biscuit, the sugar pushes the energy–protein ratio too far in the energy direction; but in the Dundee cake, rich in eggs, the balance is restored. In the batter of Yorkshire pudding, or of pancakes, or used to coat vegetables or meat in deep-fries, the balance is restored with interest. Batters deserve more attention: toad-in-the-hole, where sausages lurk in a modified Yorkshire pudding, is potentially as versatile as the pizza.

Bread, with the protein myth killed, is restored to its rightful position as the staff of life. Modern millers, who insist on removing both the bran and the wheat germ to produce the ultra-white 70 per cent sponge-rubber loaf that we are allegedly so fond of, succeed both in purging the fibre – which if left would ensure that bread need not be fattening – and in reducing the protein, mineral, and vitamin content. They do this primarily for commercial reasons which we will examine in a later chapter. In sawing through life's staff they have a lot to answer for.

In bread – including the chapatis, nans, parathas, and the rest, of India; the pitta of Greece; the tortilla of Mexico and so on – we see one of those reiterations of themes that makes the study of world cooking so fascinating. A similarly reiterated theme is the dumpling – the steamed stuffed dumplings of the wheat belt of China also flourish in various forms in eastern Europe, and in the great meat or leek puddings of the north of England, where the basic flour and water is given body by the hard fat of suet. And suet pastry – baked now, rather than steamed – is used in one of the most economical and succulent dishes of all Europe, the potato pie of Lancashire. The basic ingredients of beef, perhaps kidney, potato and onion, with loads of pepper and salt, cooked in a flower-pot shaped 'turtle' pot and left to simmer for hours in the slow oven by the perpetual coal fires of the mining towns, combine in the way that is characteristic of all classic dishes: the whole far exceeds, in flavour, the sum of its parts. Nutritionally the dish is unimpeachable. And it was born of austerity.

I doubt whether wholemeal flour, for all its nutritional and other technical advantages, should wholly replace white flour. It would be a shame to sacrifice the steamy lightness of a Chinese dumpling, for example. But whether the modern '70 per cent' flour is ever justified is a moot point: the British wartime 80–85 per cent flour, creamier in colour and with some of the bran and germ left in, seems as versatile. But a kitchen without wholemeal flour needs radical overhaul; wholemeal pastry, especially when heavily seasoned, gives an exciting edge to many a pie or pastie, while wholemeal apple crumble (the skins left on the apples and the pips left in) far outstrips the usual form. Persist with wholemeal flour; use it, and wholemeal bread if you can get it, wherever possible. It will reward you gastronomically, as well as nutritionally.

Pulses, and in particular beans, which should play such an important part in our rational agriculture and diet, are under-used and often notoriously abused by over-refined and under-informed western cooks. Nobody with a feel for cooking let alone an iota of nutritional knowledge could regard the food industry's present obsession with soya steak and other ersatz bean products with anything other than bewildered disdain. How the US, which gave us the Boston baked bean – sweetened with molasses and tomato, buffered by onion and garlic, and edged with mustard and pepper – can take ersatz seriously is almost beyond understanding. I will return to this theme later. Let us first look at what real food experts can do with beans.

According to Jane Grigson, . . . 'every civilization has its special beans. Europe's classical bean, back to the Greeks, was the broad bean. The bean of the Incas, and the Mexicans, was the kidney-bean, which Peruvian Indians were eating several thousand years ago.' The Americans can also make use of the soya, whose possibilities are already demonstrated by China and perhaps even more by Japan (though the soya steak manufacturers like to pretend the pristine bean is unpalatable), and most countries have peas and lentils.

Good cooking, traditionally, reflects the seasons: the modern obsession for food out of season, or out of place, is little more than snobbery, which is another word for philistinism. The executive who flits from four-star hotel to four-star hotel, demanding Scotch steak, French château wine, Pacific prawns, and out of season strawberries is not a gastronome, but a kind of junkie, the up-market equivalent of the Coca-Cola–peanut butter addict. His basic lack of appreciation is reflected in his dumb acceptance of over-bred, over-frozen and over-boiled vegetables. But we can follow the seasons through beans, as witness a small quote from Jane Grigson: 'Beans of all kinds are a favourite accompaniment to lamb in France. In early summer there will be a dish of the beautiful stringless haricots verts. They will be followed by shelled green flageolet beans, first fresh, then half dried as summer passes. In winter the lamb will be served on a bed of white haricot beans. Sometimes they will be cooked together . . .', and she offers a recipe with garlic, onion and carrot. There is, of course, room for lamb in rational agriculture. Especially when 'extended' – to borrow the food technologists' jargon – by beans.

Beans, like grains, may be more and more finely reduced, until you wind up with purées – a purée of lentils is a fine foil for sausages – or, more arcanely, with bessan, the bean flour of India. Usually they should not, as many food writers irritatingly insist, be soaked overnight; left so long they may begin to ferment, which spoils the flavour. Two hours is generally enough, and this may be shortened if they are brought to the boil and then left in the cooling water.

We have seen that beans, rich in lysine, complement cereal, which tends to be short of it. Obligingly, the cereal-plus-bean theme features in all the great cuisines. In China and Japan delicate beans complement rice. Dhalls, of infinite variety, are matched with chapatis in India, where bessan flour may also complement cereal flour in breads. The Mexicans wrap beans in tortillas (small boys thrived on this in John Steinbeck's *Tortilla Flat*, much to the nutritionists' astonishment) while bread and beans feature in a hundred combinations all around the Mediterranean. Beans and barley rub shoulders in many a stew in Britain, which is also the traditional home of beans on toast. I have matched haricot beans with pigeon and carrot and plenty of onion and garlic in a pie; the crust well-seasoned wholemeal. This, small fragments of wild meat extended two ways by cereal and bean is agronomic and nutritional perfection.

In the trial of traditional nutritionists, no charge is more serious than their erstwhile abuse of the potato. It should be regarded as a staple, and as a source of protein, to rank with cereal. The idea that potatoes are fattening hardly accords with the observation that protein and calorie ratio approximates to human needs: obviously you will get fat if you eat too much. Perhaps the mistake has been to take off the fibrous skin. I commend Robert Carrier's advice . . . – 'whenever possible, cook potatoes in their jackets' – and suggest that the skins can be left on more often than you think. Try cooking unpeeled but coarsely chopped potatoes in a casserole with half an inch of water in the oven while the joint cooks. Add carrot, or onion, or parsnip, or turnip, or even whole tomatoes and a half cabbage on top for good measure – and then mix the resultant vegetable juice with the meat juice for gravy, with a bit of seasoning. An ancient technique of gravy-making; in vivid contrast to the usual boiling and eleventh-hour resuscitation with proprietary gravy mix.

The potato's versatility is second only to that of cereals and no nation has employed it more ingeniously than the Swiss. One of Switzerland's many potato dishes – which I select because I think it is improved by including the skins, though most recipes recommend peeling – is rösti. One method is to boil the potatoes, let them cool, then slice them very thinly and fry in the fat of your choice, at first tossing them to get the crispy bits into the middle, and then pressing together so that the slices meld into a flat cake. Purists may question the technique, but it works, and cheese, herbs, or what you will, provide endless variations.

Swiss cuisine seems underrated, but anyone trying to get the best out of a temperate agriculture should look at it closely. . . . The Swiss use of soft fruits to make hard liquor should be studied in Britain and the US where the excursion into spirits perversely seems to begin and end with grain, potatoes and juniper berries.

What of vegetables other than pulses? I shall return to them in a later chapter. Suffice it to say that the variety that can be produced seasonally and locally is usually far greater than you imagine. Even in Britain, in April and May, the traditional months of the 'hungry gap', there are broccoli, several varieties of cabbage, spinach, late sprouts and a wide variety of roots still left in store.

In short, our rational agriculture does not require asceticism. In demanding intelligent extension of what can be produced in the home country, and of what is in season, it merely accords with the fundamental principles of good cooking. In English middle-class homes I have been expensively treated to aubergine and pepper. All very well, but would a Provençale peasant, on whose cuisine the dishes are based, use anything that did not grow in his own fields? Britain does not have the world's most favoured climate; but even in this unpromising island, we could emulate most of the great cuisines of the world, not by borrowing foreign ingredients (except for the odd spice and orange) but by borrowing only the techniques. Great cooking is eclectic in ideas. In raw materials it is chauvinistic.

You may still have doubts, I suspect, on the grounds that traditional grain-and-potato cuisines are fattening. The nutritional answer – that high-carbohydrate foods with the fibre in are a world apart from refined carbohydrate – has been covered. I should add that the task of feeding anyone who resolutely declines to take exercise, is chauffeur driven and electrically elevated from bed to office chair, without making him fat, is physiologically impossible unless he curbs his appetite by iron will.

But one of the main reasons why modern food seems so fattening is that it is designed to be eaten to excess. Specifically, we find the food industry pushing food with little bulk and high-energy content, and known to be addictive, such as sweets, Coca-Cola, sugary popcorn, and ice cream. More generally, we have evolved the idea that part of civilized living is to eat like a medieval baron every day. The modern evening meal among the middle class in Britain, and increasingly among the working class, is 'dinner' – a hot meal built around meat. The modern 'convenience food' industry is largely based on attempts to provide this unlikely meal without effort.

Yet 'dinner' is only one form among many possible. Among many sections of the British working class, the traditional evening meal is high tea. This, like all good cuisine, is infinitely variable, yet fundamentally simple. With odd bits of meat or fish or egg matched with bread and cake it is well balanced nutritionally, and has the gastronomic virtues of reflecting location and season: a Devonshire tea is different, in detail, from a Yorkshire one.

And the high tea is 'convenient'. It was traditionally produced without too much time and trouble by women who often spent their days in factory or field. But the cooking of cakes and the like could be done on one day, for the whole week; and of pies and pastries, two or three times a week; or such things could be bought ready made, from the bakery. The great 'dinner' that westerners now like to eat every day, was reserved for one or two days a week, and for feast days.

The Victorian English upper middle class invented evening dinner. It passed the long evenings and there were plenty of servants to prepare and clear up. It was good gastronomy then, because it exactly suited the people's leisured life-style. It is not appropriate to the life style of people who work all day; and the pastiches of the food industry, the not-so-instant mash and the pre-sliced beef, do not conceal its inappropriateness. A good tea is in a superior league to a bad dinner.

But the point here is that the Victorian dinner, with courses served à la Russe, was specifically designed to increase consumption. The dramatic switches from savoury to sweet and back again were a device to overcome the feelings of satiety that comes from sameness as much as from volume. You do not need ascetism to avoid overweight, unless you are physiologically aberrant; but you do need to avoid gluttony, and gluttony is built into the Victorian dinner. Simplicity is the key to good eating, except on feast days.

Indeed, I suggest that the key to all crash slimming diets is ultra-simplicity: monotony. And this applies as much to the sophisticated diets based on beefsteak, as to the old-fashioned thin toast and black coffee. In some of the modern diets you are invited to eat as much as you like: provided it is only another grapefruit, or another shive of cheese.

Incidentally, it would be good to see the restaurant more widely regarded as a place where good cooks nurtured the cuisines of their region, and where families went regularly to enjoy food and company. The most interesting restaurants in Holland, France, and Switzerland are of this kind. They are rare in the US and Britain. Too often in London the 'restaurant' is either an anonymous eating hall, for stoking up, or an expensive theatre of seduction where pretentious eclecticism fails to conceal the poverty of technique. Except for a few Chinese, Italian, and Greek restaurants, I know few where people would dream of taking their children. Sad to say, on his rare days out, junior is marshalled into the hamburger joint.

Anyway, if we could only begin to see that cooking and eating are human-sized activities, necessary but also among our most pleasurable cultural indulgences; that they should not be regarded simply as a lauching-pad for high technology and big business; then we need have no problems. We need not look to the high-flown nutritionists or faddists, and still less to the food industry, for guidance in an austere world. To misquote Winston Churchill, we merely need to put our faith in the people, for most of whom, through most of history, austerity had been a constant fact of life. . . .

Alfred Russel Wallace

'The limits of natural selection as applied to man', chapter 10 of *Contributions to the Theory of Natural Selection*, 1875

I have endeavoured to show, that the known laws of variation, multiplication, and heredity, resulting in a 'struggle for existence' and the 'survival of the fittest', have probably sufficed to produce all the varieties of structure, all the wonderful adaptations, all the beauty of form and of colour, that we see in the animal and vegetable kingdoms. To the best of my ability I have answered the most obvious and the most often repeated objections to this theory, and have, I hope, added to its general strength, by showing how colour – one of the strongholds of the advocates of special creation – may be, in almost all its modifications, accounted for by the combined influence of sexual selection and the need of protection. I have also endeavoured to show, how the same power which has modified animals has acted on man; and have, I believe, proved that, as soon as the human intellect became developed above a certain low stage, man's body would cease to be materially affected by natural selection, because the development of his mental faculties would render important modifications of its form and structure unnecessary. It will, therefore, probably excite some surprise among my readers, to find that I do not consider that all nature can be explained on the principles of which I am so ardent an advocate; and that I am now myself going to state objections, and to place limits, to the power of 'natural selection'. I believe, however, that there are such limits; and that just as surely as we can trace the action of natural laws in the development of organic forms, and can clearly conceive that fuller knowledge would enable us to follow step by step the whole process of that development, so surely can we trace the action of some unknown higher law, beyond and independent of all those laws of which we have any knowledge. We can trace this action more or less distinctly in many phenomena, the two most important of which are – the origin of sensation or consciousness, and the development of man from the lower animals. I shall first consider the latter difficulty as more immediately connected with the subjects discussed in this volume.

WHAT NATURAL SELECTION CAN NOT DO

In considering the question of the development of man by known natural laws, we must ever bear in mind the first principle of 'natural selection', no less than of the general theory of evolution, that all changes of form or structure, all increase in the size of an organ or in its complexity, all greater specialization or physiological division of labour, can only be brought about, in as much as it is for the good of the being so modified. Mr Darwin himself has taken care to impress upon us, that 'natural selection' has no power to produce absolute perfection but only relative perfection, no power to advance any being much beyond his fellow beings, but only just so much beyond them as to enable it to survive them in the struggle for existence. Still less has it any power to produce modifications which are in any degree injurious to its possessor, and Mr Darwin frequently uses the strong expression, that a single case of this kind would be fatal to his theory. If, therefore, we find in man any characters, which all the evidence we can obtain goes to show would have been actually injurious to him on their first appearance, they could not possibly have been produced by natural selection. Neither could any specially developed organ have been so produced if it had been merely useless to him, or if its use were not proportionate to its degree of development. Such cases as these would prove, that some other law, or some other power, than 'natural selection' had been at work. But if, further, we could see that these very modifications, though hurtful or useless at the time when they first appeared, became in the highest degree useful at a much later period, and are now essential to the full moral and intellectual development of human nature, we should then infer the action of mind, foreseeing the future and prepariong for it, just as surely as we do, when we see the breeder set himself to work with the determination to produce a definite improvement in some cultivated plant or domestic animal. I would further remark that this enquiry is as thoroughly scientific and legitimate as that into the origin of species itself. It is an attempt to solve the inverse problem, to deduce the existence of a new power of a definite character, in order to account for facts which according to the theory of natural selection ought not to happen. Such problems are well known to science, and the search after their solution has often led to the most brilliant results. In the case of man, there are facts of the nature above alluded to, and in calling attention to them, and in inferring a cause for them, I believe that I am as strictly within the bounds of scientific investigation as I have been in any other portion of my work.

THE BRAIN OF THE SAVAGE SHOWN TO BE LARGER THAN HE NEEDS IT
TO BE

Size of Brain an important Element of Mental Power. – The brain is universally
admitted to be the organ of the mind; and it is almost as universally
admitted, that size of brain is one of the most important of the elements
which determine mental power or capacity. There seems to be no doubt
that brains differ considerably in quality, as indicated by greater or less
complexity of the convolutions, quantity of grey matter, and perhaps
unknown peculiarities of organization; but this difference of quality seems
merely to increase or diminish the influence of quantity, not to neutralize
it. Thus, all the most eminent modern writers see an intimate connection
between the diminished size of the brain in the lower races of mankind, and
their intellectual inferiority. The collections of Dr J. B. Davis and Dr
Morton give the following as the average internal capacity of the cranium
in the chief races: – Teutonic family, 94 cubic inches; Esquimaux, 91 cubic
inches; Negroes, 85 cubic inches; Australians and Tasmanians, 82 cubic
inches; Bushmen, 77 cubic inches. These last numbers, however, are
deduced from comparatively few specimens, and may be below the average,
just as a small number of Finns and Cossacks give 98 cubic inches, or
considerably more than that of the German races. It is evident, therefore,
that the absolute bulk of the brain is not necessarily much less in savage
than in civilised man, for Esquimaux skulls are known with a capacity of
113 inches, or hardly less than the largest among Europeans. But what is
still more extraordinary, the few remains yet known of pre-historic man do
not indicate any material diminution in the size of the brain case. A Swiss
skull of the stone age, found in the lake dwelling of Meilen, corresponded
exactly to that of a Swiss youth of the present day. The celebrated
Neanderthal skull had a larger circumference than the average, and its
capacity, indicating actual mass of brain, is estimated to have been not less
than 75 cubic inches, or nearly the average of existing Australian crania.
The Engis skull, perhaps the oldest known, and which, according to Sir
John Lubbock, 'there seems no doubt was really contemporary with the
mammoth and the cave bear', is yet, according to Professor Huxley, 'a fair
average skull, which might have belonged to a philosopher, or might have
contained the thoughtless brains of a savage.' Of the cave men of Les
Eyzies, who were undoubtedly contemporary with the reindeer in the South
of France, Professor Paul Broca says (in a paper read before the Congress of
Pre-historic Archæology in 1868) – 'The great capacity of the brain, the
development of the frontal region, the fine elliptical form of the anterior
part of the profile of the skull, are incontestible characteristics of
superiority, such as we are accustomed to meet with in civilised races;' yet

the great breadth of the face, the enormous development of the ascending ramus of the lower jaw, the extent and roughness of the surfaces for the attachment of the muscles, especially of the masticators, and the extraordinary development of the ridge of the femur, indicate enormous muscular power, and the habits of a savage and brutal race.

These facts might almost make us doubt whether the size of the brain is in any direct way an index of mental power, had we not the most conclusive evidence that it is so, in the fact that, whenever an adult male European has a skull less than nineteen inches in circumference, or has less than sixty-five cubic inches of brain, he is invariably idiotic. When we join with this the equally undisputed fact, that great men – those who combine acute perception with great reflective power, strong passions, and general energy of character, such as Napoleon, Cuvier, and O'Connell, have always heads far above the average size, we must feel satisfied that volume of brain is one, and perhaps the most important, measure of intellect; and this being the case, we cannot fail to be struck with the apparent anomaly, that many of the lowest savages should have as much brains as average Europeans. The idea is suggested of a surplusage of power; of an instrument beyond the needs of its possessor.

Comparison of the Brains of Man and of Anthropoid Apes. – In order to discover if there is any foundation for this notion, let us compare the brain of man with that of animals. The adult male Orang-utan is quite as bulky as a small sized man, while the Gorilla is considerably above the average size of man, as estimated by bulk and weight; yet the former has a brain of only 28 cubic inches, the latter, one of 30, or, in the largest specimen yet known, of 34½ cubic inches. We have seen that the average cranial capacity of the lowest savages is probably not less than *five-sixths* of that of the highest civilized races, while the brain of the anthropoid apes scarcely amounts to *one-third* of that of man, in both cases taking the average; or the proportions may be more clearly represented by the following figures – anthropoid apes, 10; savages, 26; civilized man, 32. But do these figures at all approximately represent the relative intellect of the three groups? Is the savage really no further removed from the philosopher, and so much removed from the ape, as these figures would indicate? In considering this question, we must not forget that the heads of savages vary in size, almost as much as those of civilized Europeans. Thus, while the largest Teutonic skull in Dr Davis' collection is 112.4 cubic inches, there is an Araucanian of 115.5, and Esquimaux of 113.1, a Marquesan of 110.6, a Negro of 105.8, and even an Australian of 104.5 cubic inches. We may, therefore, fairly compare the savage with the highest European on the one side, and with the Orang, Chimpanzee, or Gorilla, on the other, and see whether there is any relative proportion between brain and intellect.

Range of intellectual power in Man. – First, let us consider what this wonderful instrument, the brain, is capable of in its higher developments. In Mr Galton's interesting work on 'Hereditary Genius', he remarks on the enormous difference between the intellectual power and grasp of the well-trained mathematician or man of science, and the average Englishman. The number of marks obtained by high wranglers, is often more than thirty times as great as that of the men at the bottom of the honour list, who are still of fair mathematical ability; and it is the opinion of skilled examiners, that even this does not represent the full difference of intellectual power. If, now, we descend to those savage tribes who only count to three or five, and who find it impossible to comprehend the addition of two and three without having the objects actually before them, we feel that the chasm between them and the good mathematician is so vast, that a thousand to one will probably not fully express it. Yet we know that the mass of brain might be nearly the same in both, or might not differ in a greater proportion than as 5 to 6; whence we may fairly infer that the savage possesses a brain capable, if cultivated and developed, of performing work of a kind and degree far beyond what he ever requires it to do.

Again, let us consider the power of the higher or even the average civilized man, of forming abstract ideas, and carrying on more or less complex trains of reasoning. Our languages are full of terms to express abstract conceptions. Our business and our pleasures involve the continual foresight of many contingencies. Our law, our government, and our science, continually require us to reason through a variety of complicated phenomena to the expected result. Even our games, such as chess, compel us to exercise all these faculties in a remarkable degree. Compare this with the savage languages, which contain no words for abstract conceptions; the utter want of foresight of the savage man beyond his simplest necessities; his inability to combine, or to compare, or to reason on any general subject that does not immediately appeal to his senses. So, in his moral and aesthetic faculties, the savage has none of those wide sympathies with all nature, those conceptions of the infinite, of the good, of the sublime and beautiful, which are so largely developed in civilized man. Any considerable development of these would, in fact, be useless or even hurtful to him, since they would to some extent interfere with the supremacy of those perceptive and animal faculties on which his very existence often depends, in the severe struggle he has to carry on against nature and his fellow-man. Yet the rudiments of all these powers and feelings undoubtedly exist in him, since one or other of them frequently manifest themselves in exceptional cases, or when some special circumstances call them forth. Some tribes, such as the Santals, are remarkable for as pure a love of truth as the most moral among civilized men. The . . . Polynesian . . . [has] a high artistic feeling, the first traces of which are clearly visible in the rude drawings of the palæolithic men who were the contemporaries in France of the Reindeer and the Mammoth. Instances of unselfish love, of true gratitude, and of deep religious feeling, sometimes occur among most savage races.

On the whole, then, we may conclude, that the general moral and intellectual development of the savage, is not less removed from that of civilized man than has been shown to be the case in the one department of mathematics; and from the fact that all the moral and intellectual faculties do occasionally manifest themselves, we may fairly·conclude that they are always latent, and that the large brain of the savage man is much beyond his actual requirements in the savage state.

Intellect of Savages and of Animals compared. – Let us now compare the intellectual wants of the savage, and the actual amount of intellect he exhibits, with those of the higher animals. Such races as the Andaman Islanders, the Australians, and the Tasmanians, the Digger Indians of North America, or the natives of Fuegia, pass their lives so as to require the exercise of few faculties not possessed in an equal degree by many animals. In the mode of capture of game or fish, they by no means surpass the ingenuity or forethought of the jaguar, who drops saliva into the water, and seizes the fish as they come to eat it; or of wolves and jackals, who hunt in packs; or of the fox, who buries his surplus food till he requires it. The sentinels placed by antelopes and by monkeys, and the various modes of building adopted by field mice and beavers, as well as the sleeping place of the orang-utan, and the tree-shelter of some of the African anthropoid apes, may well be compared with the amount of care and forethought bestowed by many savages in similar circumstances. His possession of free and perfect hands, not required for locomotion, enable man to form and use weapons and implements which are beyond the physical powers of brutes; but having done this, he certainly does not exhibit more mind in using them than do many lower animals. What is there in the life of the savage, but the satisfying of the cravings of appetite in the simplest and easiest way? What thoughts, ideas, or actions are there, that raise him many grades above the elephant or the ape? Yet he possesses, as we have seen, a brain vastly superior to theirs in size and complexity; and this brain gives him, in an undeveloped state, faculties which he never requires to use. And if this is true of existing savages, how much more true must it have been of the men whose sole weapons were rudely chipped flints, and some of whom, we may fairly conclude, were lower than any existing race; while the only evidence yet in our possession shows them to have had brains fully as capacious as those of the average of the lower savage races.

We see, then, that whether we compare the savage with the higher developments of man, or with the brutes around him, we are alike driven to the conclusion that in his large and well-developed brain he possesses an organ quite disproportionate to his actual requirements – an organ that seems prepared in advance, only to be fully utilized as he progresses in civilization. A brain slightly larger than that of the gorilla would, according to the evidence before us, fully have sufficed for the limited mental development of the savage; and we must therefore admit, that the large brain he actually possesses could never have been solely developed by any of those laws of evolution, whose essence is, that they lead to a degree of organization exactly proportionate to the wants of each species, never beyond those wants – that no preparation can be made for the future development of the race – that one part of the body can never increase in size or complexity, except in strict co-ordination to the pressing wants of the whole. The brain of pre-historic and of savage man seems to me to prove the existence of some power, distinct from that which has guided the development of the lower animals through their ever-varying forms of being.

THE USE OF THE HAIRY COVERING OF MAMMALIA

Let us now consider another point in man's organization, the bearing of which has been almost entirely overlooked by writers on both sides of this question. One of the most general external characters of the terrestrial mammalia is the hairy covering of the body, which, whenever the skin is flexible, soft, and sensitive, forms a natural protection against the severities of climate, and particularly against rain. That this is its most important function, is well shown by the manner in which the hairs are disposed so as to carry off the water, by being invariably directed downwards from the most elevated parts of the body. Thus, on the under surface the hair is always less plentiful, and, in many cases, the belly is almost bare. The hair lies downwards, on the limbs of all walking mammals, from the shoulder to the toes, but in the orang-utan it is directed from the shoulder to the elbow, and again from the wrist to the elbow, in a reverse direction. This corresponds to the habits of the animal, which, when resting, holds its long arms upwards over its head, or clasping a branch above it, so that the rain would flow down both the arm and fore-arm to the long hair which meets at the elbow. In accordance with this principle, the hair is always longer or more dense along the spine or middle of the back from the nape to the tail, often rising into a crest of hair or bristles on the ridge of the back. This character prevails through the entire series of the mammalia, from the marsupials to the quadrumana, and by this long persistence it must have acquired such a powerful hereditary tendency, that we should expect it to reappear continually even after it had been abolished by ages of the most rigid selection; and we may feel sure that it never could have been completely abolished under the law of natural selection, unless it had become so positively injurious as to lead to the almost invariable extinction of individuals possessing it.

THE CONSTANT ABSENCE OF HAIR FROM CERTAIN PARTS OF MAN'S BODY A REMARKABLE PHENOMENON

In man the hairy covering of the body has almost totally disappeared, and, what is very remarkable, it has disappeared more completely from the back than from any other part of the body. Bearded and beardless races alike have the back smooth, and even when a considerable quantity of hair appears on the limbs and breast, the back, and especially the spinal region, is absolutely free, thus completely reversing the characteristics of all other mammalia. The Ainos of the Kurile Islands and Japan are said to be a hairy race; but Mr Bickmore, who saw some of them, and described them in a paper read before the Ethnological Society, gives no details as to where the hair was most abundant, merely stating generally, that 'their chief peculiarity is their great abundance of hair, not only on the head and face, but over the whole body.' This might very well be said of any man who had hairy limbs and breast, unless it was specially stated that his back was hairy, which is not done in this case. The hairy family in Birmah have, indeed, hair on the back rather longer than on the breast, thus reproducing the true mammalian character, but they have still longer hair on the face, forehead, and inside the ears, which is quite abnormal; and the fact that their teeth are all very imperfect, shows that this is a case of monstrosity rather than one of true reversion to the ancestral type of man before he lost his hairy covering.

SAVAGE MAN FEELS THE WANT OF THIS HAIRY COVERING

We must now enquire if we have any evidence to show, or any reason to believe, that a hairy covering to the back would be in any degree hurtful to savage man, or to man in any stage of his progress from his lower animal form; and if it were merely useless, could it have been so entirely and completely removed as not to be continually reappearing in mixed races? Let us look to savage man for some light on these points. One of the most common habits of savages is to use some covering for the back and shoulders, even when they have none on any other part of the body. The early voyagers observed with surprise, that the Tasmanians, both men and women, wore the kangaroo-skin, which was their only covering, not from any feeling of modesty, but over the shoulders to keep the back dry and warm. A cloth over the shoulders was also the national dress of the Maories. The Patagonians wear a cloak or mantle over the shoulders, and the Fuegians often wear a small piece of skin on the back, laced on, and shifted from side to side as the wind blows. The Hottentots also wore a somewhat similar skin over the back, which they never removed, and in which they were buried. Even in the tropics most savages take precautions to keep their backs dry. The natives of Timor use the leaf of a fan palm, carefully stitched up and folded, which they always carry with them, and which, held over the back, forms an admirable protection from the rain. Almost all the Malay races, as well as the Indians of South America, make great palm-leaf hats, four feet or more across, which they use during their canoe voyages to protect their bodies from heavy showers of rain; and they use smaller hats of the same kind when travelling by land.

We find, then, that so far from there being any reason to believe that a hairy covering to the back could have been hurtful or even useless to pre-historic man, the habits of modern savages indicate exactly the opposite view, as they evidently feel the want of it, and are obliged to provide substitutes of various kinds. The perfectly erect posture of man, may be supposed to have something to do with the disappearance of the hair from his body, while it remains on his head; but when walking, exposed to rain and wind, a man naturally stoops forwards, and thus exposes his back; and the undoubted fact, that most savages feel the effects of cold and wet most severely in that part of the body, sufficiently demonstrates that the hair could not have ceased to grow there merely because it was useless, even if it were likely that a character so long persistent in the entire order of mammalia, could have so completely disappeared, under the influence of so weak a selective power as a diminished usefulness.

MAN'S NAKED SKIN COULD NOT HAVE BEEN PRODUCED BY NATURAL SELECTION

It seems to me, then, to be absolutely certain, that 'Natural Selection' could not have produced man's hairless body by the accumulation of variations from a hairy ancestor. The evidence all goes to show that such variations could not have been useful, but must, on the contrary, have been to some extent hurtful. If even, owing to an unknown correlation with other hurtful qualities, it had been abolished in the ancestral tropical man, we cannot conceive that, as man spread into colder climates, it should not have returned under the powerful influence of reversion to such a long persistent ancestral type. But the very foundation of such a supposition as this is untenable; for we cannot suppose that a character which, like hairiness, exists throughout the whole of the mammalia, can have become, in one form only, so constantly correlated with an injurious character, as to lead to its permanent suppression – a suppression so complete and effectual that it never, or scarcely ever, reappears in mongrels of the most widely different races of man.

Two characters could hardly be wider apart, than the size and development of man's brain, and the distribution of hair upon the surface of his body; yet they both lead us to the same conclusion – that some other power than Natural Selection has been engaged in his production.

FEET AND HANDS OF MAN, CONSIDERED AS DIFFICULTIES ON THE THEORY OF NATURAL SELECTION

There are a few other physical characteristics of man, that may just be mentioned as offering similar difficulties, though I do not attach the same importance to them as to those I have already dwelt on. The specialization and perfection of the hands and feet of man seems difficult to account for. Throughout the whole of the quadrumana the foot is prehensile; and a very rigid selection must therefore have been needed to bring about that arrangement of the bones and muscles, which has converted the thumb into a great toe, so completely, that the power of opposability is totally lost in every race, whatever some travellers may vaguely assert to the contrary. It is difficult to see why the prehensile power should have been taken away. It must certainly have been useful in climbing, and the case of the baboons shows that it is quite compatible with terrestrial locomotion. It may not be compatible with perfectly easy erect locomotion; but, then, how can we conceive that early man, *as an animal*, gained anything by purely erect locomotion? Again, the hand of man contains latent capacities and powers which are unused by savages, and must have been even less used by palæolithic man and his still ruder predecessors. It has all the appearance of an organ prepared for the use of civilized man, and one which was required to render civilization possible. Apes make little use of their separate fingers and opposable thumbs. They grasp objects rudely and clumsily, and look as if a much less specialized extremity would have served their purpose as well. I do not lay much stress on this, but, if it be proved that some intelligent power has guided or determined the development of man, then we may see indications of that power, in facts which, by themselves, would not serve to prove its existence.

The voice of man. – The same remark will apply to another peculiarly human character, the wonderful power range, flexibility, and sweetness, of the musical sounds producible by the human larynx, especially in the female sex. The habits of savages give no indication of how this faculty could have been developed by natural selection; because it is never required or used by them. The singing of savages is a more or less monotonous howling, and the females seldom sing at all. Savages certainly never choose their wives for fine voices, but for rude health, and strength, and physical beauty. Sexual selection could not therefore have developed this wonderful power, which only comes into play among civilized people. It seems as if the organ had been prepared in anticipation of the future progress of man, since it contains latent capacities which are useless to him in his earlier condition. The delicate correlations of structure that give it such marvellous powers, could not therefore have been acquired by means of natural selection.

THE ORIGIN OF SOME OF MAN'S MENTAL FACULTIES, BY THE PRESERVATION OF USEFUL VARIATIONS, NOT POSSIBLE

Turning to the mind of man, we meet with many difficulties in attempting to understand, how those mental faculties, which are especially human, could have been acquired by the preservation of useful variations. At first sight, it would seem that such feelings as those of abstract justice and benevolence could never have been so acquired because they are incompatible with the law of the strongest, which is the essence of natural selection. But this is, I think, an erroneous view, because we must look, not to individuals but to societies; and justice and benevolence, exercised towards members of the same tribe, would certainly tend to strengthen that tribe, and give it a superiority over another in which the right of the strongest prevailed, and where consequently the weak and the sickly were left to perish, and the few strong ruthlessly destroyed the many who were weaker.

But there is another class of human faculties that do not regard our fellow men, and which cannot, therefore, be thus accounted for. Such are the capacity to form ideal conceptions of space and time, of eternity and infinity – the capacity for intense artistic feelings of pleasure, in form, colour, and composition – and for those abstract notions of form and number which render geometry and arithmetic possible. How were all or any of these faculties first developed, when they could have been of no possible use to man in his early stages of barbarism? How could 'natural selection', or survival of the fittest in the struggle for existence, at all favour the development of mental powers so entirely removed from the material necessities of savage men, and which even now, with our comparatively high civilization, are, in their farthest developments, in advance of the age, and appear to have relation rather to the future of the race than to its actual status?

DIFFICULTY AS TO THE ORIGIN OF THE MORAL SENSE

Exactly the same difficulty arises, when we endeavour to account for the development of the moral sense or conscience in savage man; for although the *practice* of benevolence, honesty, or truth, may have been useful to the tribe possessing these virtues, that does not at all account for the peculiar *sanctity*, attached to actions which each tribe considers right and moral, as contrasted with the very different feelings with which they regard what is merely *useful*. The utilitarian hypothesis (which is the theory of natural selection applied to the mind) seems inadequate to account for the development of the moral sense. This subject has been recently much discussed, and I will here only give one example to illustrate my argument. The utilitarian sanction for truthfulness is by no means very powerful or universal. Few laws enforce it. No very severe reprobation follows untruthfulness. In all ages and countries, falsehood has been thought allowable in love, and laudable in war; while, at the present day, it is held to be venial by the majority of mankind, in trade, commerce, and speculation. A certain amount of untruthfulness is a necessary part of politeness in the east and west alike, while even severe moralists have held a lie justifiable, to elude an enemy or prevent a crime. Such being the difficulties with which this virtue has had to struggle, with so many exceptions to its practice, with so many instances in which it brought ruin or death to its too ardent devotee, how can we believe that considerations of utility could ever invest it with the mysterious sanctity of the highest virtue, – could ever induce men to value truth for its own sake, and practice it regardless of consequences?

Yet, it is a fact, that such a mystical sense of wrong does attach to untruthfulness, not only among the higher classes of civilized people, but among whole tribes of utter savages. Sir Walter Elliott tells us (in his paper 'On the Characteristics of the Population of Central and Southern India', published in the Journal of the Ethnological Society of London, vol. i., p. 107) that the Kurubars and Santals, barbarous hill-tribes of Central India, are noted for veracity. It is a common saying that 'a Kurubar *always* speaks the truth', and Major Jervis says, 'the Santals are the most truthful men I ever met with.' As a remarkable instance of this quality the following fact is given. A number of prisoners, taken during the Santal insurrection, were allowed to go free on parole, to work at a certain spot for wages. After some time cholera attacked them and they were obliged to leave, but every man of them returned and gave up his earnings to the guard. Two hundred savages with money in their girdles, walked thirty miles back to prison rather than break their word! My own experience among savages has furnished me with similar, although less severely tested, instances; and we cannot avoid asking, how is it, that in these few cases 'experiences of utility' have left such an overwhelming impression, while in so many others they have left none? The experiences of savage men as regards the utility of truth, must, in the long run, be pretty nearly equal. How is it, then, that in some cases the result is a sanctity which overrides all considerations of personal advantage, while in others there is hardly a rudiment of such a feeling?

The intuitional theory, which I am now advocating, explains this by the supposition, that there is a feeling – a sense of right and wrong – in our nature, antecedent to and independent of experiences of utility. Where free play is allowed to the relations between man and man, this feeling attaches itself to those acts of universal utility or self-sacrifice, which are the products of our affections and sympathies, and which we term moral; while it may be, and often is, perverted, to give the same sanction to acts of narrow and conventional utility which are really immoral. . . .

The strength of the moral feeling will depend upon individual or racial constitution, and on education and habit; – the acts to which its sanctions are applied, will depend upon how far the simple feelings and affections of our nature, have been modified by custom, by law, or by religion.

It is difficult to conceive that such an intense and mystical feeling of right and wrong, (so intense as to overcome all ideas of personal advantage or utility), could have been developed out of accumulated ancestral experiences of utility; and still more difficult to understand, how feelings developed by one set of utilities, could be transferred to acts of which the utility was partial, imaginary, or altogether absent. But if a moral sense is an essential part of our nature, it is easy to see, that its sanction may often be given to acts which are useless or immoral; just as the natural appetite for drink, is perverted by the drunkard into the means of his destruction.

SUMMARY OF THE ARGUMENT AS TO THE INSUFFICIENCY OF NATURAL SELECTION TO ACCOUNT FOR THE DEVELOPMENT OF MAN

Briefly to resume my argument – I have shown that the brain of the lowest savages, and, as far as we yet know, of the pre-historic races, is little inferior in size to that of the highest types of man, and immensely superior to that of the higher animals; while it is universally admitted that quantity of brain is one of the most important, and probably the most essential, of the elements which determine mental power. Yet the mental requirements of savages, and the faculties actually exercised by them, are very little above those of animals. The higher feelings of pure morality and refined emotion, and the power of abstract reasoning and ideal conception, are useless to them, are rarely if ever manifested, and have no important relations to their habits, wants, desires, or well-being. They possess a mental organ beyond their needs. Natural Selection could only have endowed savage man with a brain a little superior to that of an ape, whereas he actually possesses one very little inferior to that of a philosopher.

The soft, naked, sensitive skin of man, entirely free from that hairy covering which is so universal among other mammals, cannot be explained on the theory of natural selection. The habits of savages show that they feel the want of this covering, which is most completely absent in man exactly where it is thickest in other animals. We have no reason whatever to believe, that it could have been hurtful, or even useless to primitive man; and, under these circumstances, its complete abolition, shown by its never reverting in mixed breeds, is a demonstration of the agency of some other power than the law of the survival of the fittest, in the development of man from the lower animals.

Other characters show difficulties of a similar kind, though not perhaps in an equal degree. The structure of the human foot and hand seem unnecessarily perfect for the needs of savage man, in whom they are as completely and as humanly developed as in the highest races. The structure of the human larynx, giving the power of speech and of producing musical sounds, and especially its extreme development in the female sex, are shown to be beyond the needs of savages, and from their known habits, impossible to have been acquired either by sexual selection, or by survival of the fittest.

The mind of man offers arguments in the same direction, hardly less strong than those derived from his bodily structure. A number of his mental faculties have no relation to his fellow men, or to his material progress. The power of conceiving eternity and infinity, and all those purely abstract notions of form, number, and harmony, which play so large a part in the life of civilised races, are entirely outside of the world of thought of the savage, and have no influence on his individual existence or on that of his tribe. They could not, therefore, have been developed by any preservation of useful forms of thought; yet we find occasional traces of them amidst a low civilization, and at a time when they could have had no practical effect on the success of the individual, the family, or the race; and the development of a moral sense or conscience by similar means is equally inconceivable.

But, on the other hand, we find that every one of these characteristics is necessary for the full development of human nature. The rapid progress of civilization under favourable conditions, would not be possible, were not the organ of the mind of man prepared in advance, fully developed as regards size, structure, and proportions, and only needing a few generations of use and habit to co-ordinate its complex functions. The naked and sensitive skin, by necessitating clothing and houses, would lead to the more rapid development of man's inventive and constructive faculties; and, by leading to a more refined feeling of personal modesty, may have influenced, to a considerable extent, his moral nature. The erect form of man, by freeing the hands from all locomotive uses, has been necessary for his intellectual advancement; and the extreme perfection of his hands, has alone rendered possible that excellence in all the arts of civilization which raises him so far above the savage, and is perhaps but the forerunner of a higher intellectual and moral advancement. The perfection of his vocal organs has first led to the formation of articulate speech, and then to the development of those exquisitely toned sounds, which are only appreciated by the higher races, and which are probably destined for more elevated uses and more refined enjoyment, in a higher condition than we have yet attained to. So, those faculties which enable us to transcend time and space, and to realize the wonderful conceptions of mathematics and philosophy, or which give us an intense yearning for abstract truth (all of which were occasionally manifested at such an early period of human history as to be far in advance of any of the few practical applications which have since grown out of them), are evidently essential to the perfect development of man as a spiritual being, but are utterly inconceivable as having been produced through the action of a law which looks only, and can look only, to the immediate material welfare of the individual or the race.

The inference I would draw from this class of phenomena is, that a superior intelligence has guided the development of man in a definite direction, and for a special purpose, just as man guides the development of many animal and vegetable forms. The laws of evolution alone would, perhaps, never have produced a grain so well adapted to man's use as wheat and maize; such fruits as the seedless banana and bread-fruit; or such animals as the Guernsey milch cow, or the London dray-horse. Yet these so closely resemble the unaided productions of nature, that we may well imagine a being who had mastered the laws of development of organic forms through past ages, refusing to believe that any new power had been concerned in their production, and scornfully rejecting the theory (as my theory will be rejected by many who agree with me on other points), that in these few cases a controlling intelligence had directed the action of the laws of variation, multiplication, and survival, for his own purposes. We know, however, that this has been done; and we must therefore admit the possibility that, if we are not the highest intelligences in the universe, some higher intelligence may have directed the process by which the human race was developed, by means of more subtle agencies than we are acquainted with. At the same time I must confess, that this theory has the disadvantage of requiring the intervention of some distinct individual intelligence, to aid in the production of what we can hardly avoid considering as the ultimate aim and outcome of all organized existence – intellectual, ever-advancing, spiritual man. It therefore implies, that the great laws which govern the material universe were insufficient for his production, unless we consider (as we may fairly do) that the controlling action of such higher intelligences is a necessary part of those laws, just as the action of all surrounding organisms is one of the agencies in organic development. But even if my particular view should not be the true one, the difficulties I have put forward remain, and I think prove, that some more general and more fundamental law underlies that of 'natural selection'. The law of 'unconscious intelligence' pervading all organic nature, put forth by Dr Laycock and adopted by Mr Murphy, is such a law; but to my mind it has the double disadvantage of being both unintelligible and incapable of any kind of proof. It is more probable, that the true law lies too deep for us to discover it; but there seems to me, to be ample indications that such a law does exist, and is probably connected with the absolute origin of life and organization. . . .

THE ORIGIN OF CONSCIOUSNESS

The question of the origin of sensation and of thought can be but briefly discussed in this place, since it is a subject wide enough to require a separate volume for its proper treatment. No physiologist or philosopher has yet ventured to propound an intelligible theory, of how sensation may possibly be a product of organization; while many have declared the passage from matter to mind to be inconceivable. In his presidential address to the Physical Section of the British Association at Norwich, in 1868, Professor Tyndall expressed himself as follows: –

'The passage from the physics of the brain to the corresponding facts of consciousness is unthinkable. Granted that a definite thought, and a definite molecular action in the brain occur simultaneously, we do not possess the intellectual organ, nor apparently any rudiment of the organ, which would enable us to pass by a process of reasoning from the one phenomenon to the other. They appear together, but we do not know why.

Were our minds and senses so expanded, strengthened, and illuminated as to enable us to see and feel the very molecules of the brain; were we capable of following all their motions, all their groupings, all their electric discharges, if such there be, and were we intimately acquainted with the corresponding states of thought and feeling, we should be as far as ever from the solution of the problem, "How are these physical processes connected with the facts of consciousness?" The chasm between the two classes of phenomena would still remain intellectually impassable.'

In his latest work (*An Introduction to the Classification of Animals*), published in 1869, Professor Huxley unhesitatingly adopts the 'well founded doctrine, that life is the cause and not the consequence of organization.' In his celebrated article 'On the Physical Basis of Life', however, he maintains, that life is a property of protoplasm, and that protoplasm owes its properties to the nature and disposition of its molecules. Hence he terms it 'the matter of life', and believes that all the physical properties of organized beings are due to the physical properties of protoplasm. So far we might, perhaps, follow him, but he does not stop here. He proceeds to bridge over that chasm which Professor Tyndall has declared to be 'intellectually impassable', and, by means which he states to be logical, arrives at the conclusion, that our '*thoughts are the expression of molecular changes in that matter of life which is the source of our other vital phenomena.*' Not having been able to find any clue in Professor Huxley's writings, to the steps by which he passes from those vital phenomena, which consist only, in their last analysis, of movements of particles of matter, to those other phenomena which we term thought, sensation, or consciousness; but, knowing that so positive an expression of opinion from him will have great weight with many persons, I shall endeavour to show, with as much brevity as is compatible with clearness, that this theory is not only incapable of proof, but is also, as it appears to me, inconsistent with accurate conceptions of molecular physics. To do this, and in order further to develop my views, I shall have to give a brief sketch of the most recent speculations and discoveries, as to the ultimate nature and constitution of matter.

H. G. Wells

The opening paragraph of *The War of the Worlds*, 1898

No one would have believed, in the last years of the nineteenth century, that human affairs were being watched keenly and closely by intelligences greater than man's and yet as moral as his own; that as men busied themselves about their affairs they were scrutinized and studied, perhaps almost as narrowly as a man with a microscope might scrutinize the transient creatures that swarm and multiply in a drop of water. With infinite complacency men went to and fro over this globe about their little affairs, serene in their assurance of their empire over matter. It is possible that the infusoria under the microscope do the same. No one gave a thought to the older worlds of space as sources of human danger, or thought of them only to dismiss the idea of life upon them as impossible or improbable. It is curious to recall some of the mental habits of those departed days. At most, terrestrial men fancied there might be other men upon Mars, perhaps inferior to themselves and ready to welcome a missionary enterprise. Yet, across the gulf of space, minds that are to our minds as ours are to those of the beasts that perish, intellects vast and cool and unsympathetic, regarded this earth with envious eyes, and slowly and surely drew their plans against us. And early in the twentieth century came the great disillusionment.

An extract from *Experiment in Autobiography*, 1934

It would seem that Professor Guthrie, while he was incubating this course, had been impressed with the idea that most of his students were destined to be teachers or experimental workers and that they would find themselves in need of apparatus. Unaware of the economic forces that evoke supply in response to demand, he decided that it was a matter of primary necessity that we should learn to make that apparatus for ourselves. Then even upon desert islands or in savage jungles we should not be at a loss if suddenly an evening class surrounded us. Accordingly he concentrated our energies upon apparatus making. He swept aside the idea that physics is an experimental science and substituted a confused workshop training. When I had gone into the zoological laboratory upstairs, I had been confronted by a newly killed rabbit; I had began forthwith upon its dissection and in a week or so I had acquired a precise and ample knowledge of mammalian anatomy up to and including the structure of the brain, based upon my dissections and drawings and a careful comparison with prepared dissections of other types. Now when I came into the physics laboratory I was given a blowpipe, a piece of glass tubing, a slab of wood which required planing and some bits of paper and brass, and I was told I had to make a barometer. So instead of a student I became an amateur glass worker and carpenter.

After breaking a fair amount of glass and burning my fingers severely several times, I succeeded in sealing a yard's length tube, bending it, opening out the other end, tacking it on to the plank, filling it with mercury, attaching a scale to it and producing the most inelegant and untruthful barometer the world has ever seen. In the course of some days of heated and uncongenial effort, I had learnt nothing about the barometer, atmospheric pressure, or the science of physics that I had not known thoroughly before I left Midhurst, unless it was the blistering truth that glass can still be intensely hot after it has ceased to glow red.

I was then given a slip of glass on which to etch a millimetre scale with fluorine. Never had millimetre intervals greater individuality than I gave to mine. Again I added nothing to my knowledge – and I stained my only pair of trousers badly with acid. . . .

A wiser and more determined character than I, might have held firmly to my initial desire to learn and know about this moving framework of matter in which life is set, might have sought out books and original literature, acquired whatever mathematical equipment was necessary, and come round behind the slow obstructive Guthrie and the swift elusive Boys, outflanking them so to speak, and getting to the citadel, if any, at the centre of the thickets and wildernesses of knowledge they were failing to guide me through. I did not realize it then, but at that time the science of physics was in a state of confusion and reconstruction, and lucid expositions of the new ideas for the student and the general reader did not exist. Quite apart from its unsubstantial equipment and the lack of time, my mind had not the strength and calibre to do so much original exploration as was needed to get near to what was going on. I made a kind of effort to formulate and approach these primary questions, but my effort was not sustained.

In the students' Debating Society, of which I will tell more later, I heard about and laid hold of the idea of a four dimensional frame for a fresh apprehension of physical phenomena, which afterwards led me to send a paper, 'The Universe Rigid', to the *Fortnightly Review* (a paper which was rejected by Frank Harris as incomprehensible), and gave me a frame for my first scientific fantasia, the *Time Machine*, and there was moreover a rather elaborate joke going on with Jennings and the others, about a certain 'Universal Diagram' I proposed to make, from which all phenomena would be derived by a process of deduction. (One began with a uniformly distributed ether in the infinite space of those days and then displaced a particle. If there was a Universe rigid, and hitherto uniform, the character of the consequent world would depend entirely, I argued along strictly materialist lines, upon the velocity of this initial displacement. The disturbance would spread outward with ever increasing complication. But I discovered no way, and there was no one to show me a way to get on from such elementary struggles with primary concepts, to a sound understanding of contemporary experimental physics.)

Failing that, my mind relapsed into that natural protest of the frustrated – malicious derision of the physics presented to us. I set myself to guy and contemn Guthrie's instructions in every possible way, I took to absenting myself from the laboratory and when I was recalled to my attendances by the registrar of the schools, I brought in Latin and German textbooks and studied them ostentatiously. In those days the matriculation examination of the London University was open to all comers; it was a discursive examination involving among other things a superficial knowledge of French, Latin and either German or Greek and I found German the easier alternative. I mugged it up for myself to the not very exacting standard required. I matriculated in January 1886 as a sort of demonstration of the insufficiency of the physics course to occupy my mind.

My campaign to burlesque Guthrie's practical work was not a very successful one, it was a feeble rebellion with the odds all against me, but it amused some of my fellow students and made me some friends. Even had I been trying to satisfy the requirements of the course, the inattentive clumsiness that had already made me a failure as a shop assistant, would have introduced an element of absurdity into the barometers, thermo- meters, galvanometers, demonstration apparatus and so forth that I manufactured, but I added to this by demanding a sound scientific reason for every detail in the instructions given me and contriving some other, and usually grotesque, way of achieving the required result if such an imperative reason was not forthcoming. The laboratory instructor Mitchell was not a very quick-minded or intelligent man, bad at an argument and rather disposed to make a meticulous adhesion to instructions a matter of discipline. That gave me a great advantage over him because his powers of enforcement were strictly limited. After a time he began to avoid my end of the laboratory and when he found my bench littered with bits of stuff, a scamped induction coil or suchlike object in a state of scandalous incompleteness and myself away, he thanked his private gods and no longer reported my absence.

The decisive struggle which persuaded him to despair of me, turned upon the measurement of the vibrations of a tuning-fork giving the middle C of an ordinary piano. We had to erect a wooden cross on a stand with pins at the ends of the arms, and a glass plate, carefully blackened with candle smoke, was hung by a piece of silk passing over these arms in such a way as just to touch a bristle attached to a tuning-fork. This tuning-fork was thrown into sympathetic vibration by another, the silk thread was burnt in the middle, the plate as it fell rubbed against the bristle and a trace of the vibrations was obtained. A careful measurement of this trace and a fairly simple calculation (neglecting the buoyant effect of the atmosphere) gave the rate of vibration per second. I objected firstly to the neglect of the atmospheric resistance and I tried to worry Mitchell into some definite statement of the extent to which it vitiated the precision of the experiment. Poor dear! all that he could say was that it 'didn't amount to much'. But we joined issue more seriously upon the cross-piece. I alleged that as a non-Christian I objected to making a cross if that was avoidable. I declared that as a Deist I would prefer to hang my falling plate from one single pin. Also I insisted that it was the duty of a scientific worker always to take the simplest course to his objective. This cross-piece with its two pins was, I argued, a needless elaboration probably tainted by the theological prepossessions of Professor Guthrie. In fact I refused to make it. I could get just as good results with a Monotheistic upright. Mitchell fell into the trap by insisting that that was 'how it had to be done'. Whereupon I asked whether I was a student of physical science or a convict under discipline. Was I there to learn or was I there to obey?

Obviously Mitchell had no case and as obviously I was making a confounded nuisance of myself for no visible reason. He was acting under direction. My retrospective sympathies are entirely with him.

One example is as good as a score of the silly bickering resistances I put up to annoy my teachers during that futile course of instruction. In the end when my apparatus was assembled for inspection and marking, it was of such a distinguished badness that it drew an admiring group of fellow students and some of it was preserved in a cupboard for several years. As a comment on Professor Guthrie's conception of education it was worth preserving. But I pretended to be prouder of that collection than in my heart I was. Guthrie was taking life at an angle different from mine and I had been betrayed into some very ungracious and insulting reactions. Poor discipline goes with poor teaching. A lecture theatre full of impatient undergraduate students is the least likely of any audience to detect the presence of failing health. His husky voice strained against our insurgent hum. He was irritable and easily 'drawn'. There was a considerable amount of ironical applause and petty rowdiness during his lectures and in these disturbances I had made myself conspicuous.

Denys Wilkinson

'The organization of the universe', Wolfson College Lecture, 1980

Curiosity about the influences that shape the Universe of which we are part, about its fine structure and the ultimate constituent particles of matter, if such exist, is not by any means new in the thoughts of Man. The so-called atomists of the fifth century B.C., Leucippus and Democritus, had ideas that, cast into modern idiom, we could recognize as the germs of the way in which we now think about matter. But it is with Epicurus, as relayed to us from 'that peaceful garden' by Lucretius, that we first find a grasp of the idea that the apparent complexity and infinite variety of the visible world may hide an inner simplicity of structure existing at levels beneath our perception. Epicurus lived between the fourth and third centuries; Lucretius must have lived until at least 55 B.C., because he wrote disparagingly of the English climate in his great Epicurean work *De Rerum Natura*, which we might otherwise consider as an objective reflection upon Nature, morals, and the senses. I shall look to Lucretius for occasional comments upon my story and cannot do better than to begin by bidding you, through him:

> Give your mind now to the true reasoning I have to unfold. A new fact is battling strenuously for access to your ears. A new aspect of the Universe is striving to reveal itself. But no fact is so simple that it is not harder to believe than to doubt at the first presentation.

FORCES

The influences that shape our Universe are: the particles out of which our world is made, the forces between those particles, and the laws governing the play of those forces. We now recognize, phenomenologically, in our present-day Universe, four apparently distinct forces:

(a) *gravity*, which acts indiscriminately and attractively between all particles;

(b) the electric force that acts by definition only between electrically charged particles and that can be attractive or repulsive depending on whether the particles in question have opposite or similar charges; with electricity goes magnetism, generated by moving electric charges, as we have understood since the days of Faraday and Maxwell over a century ago, and the two are collectively called *electromagnetism*;

(c) the *strong*, or nuclear, force that acts only within a class of particles called hadrons, of which the neutrons and protons out of which the nucleus of the atom is made are examples, and which is indiscriminate in respect of the electrical charges of those particles;

(d) the *weak* force, which acts within another class of particles called leptons, of which the electron is an example; its action is independent of electric charge but it also acts between leptons and hadrons, and between hadrons and hadrons.

These four forces of Nature differ also in some of the laws that they obey: all the forces appear to conserve energy and momentum for example, but whereas the electromagnetic and strong forces show no intrinsic preference as between left-handed and right-handed phenomena (phenomena of a corkscrew nature such as circularly-polarized light or the spinning of a particle with respect to its direction of motion) the weak force does not respect this plausible symmetry; it shows a strong prejudice and can carry out certain operations only right-handedly and others only left-handedly. Thus in the process of radioactive beta-decay the electron that is emitted spontaneously from a neutron under the influence of the weak force, converting the neutron into a proton, comes out preferentially as a left-handed corkscrew; that is to say, as it departs it is spinning anti-clockwise as viewed by the proton that it leaves behind. The concept of symmetry and the behaviour of the forces under different kinds of symmetry transformation, such as replacing left-handedly spinning particles with right-handed ones, are of fundamental significance, as Chris Llewellyn Smith explains in chapter 3.

The four forces also differ in the way in which their effect decreases as the distance between the particles between which they act increases: the gravitational and electrical forces drop off only rather slowly with separation – the famous inverse square law of force – whereas the strong force falls off much more sharply than this and has a reach of only about 10^{-13} cm, beyond which it very rapidly becomes negligible; the weak force is also of short range in this same sense – we believe that its reach is very much less even than that of the strong force. These differences in range and strength may be illustrated by considering the force between two hydrogen atoms as a function of their distance apart. Each atom is a proton with an electron going around it at a distance of about 1 Å (or 10^{-8} cm) and we will take as a measure of separation the distance between the protons at their centres. At a separation of 1 fm (or 10^{-13} cm), a distance characteristic of the finite reach of the strong force, the strong force is about 100 times stronger than the electric force. The weak and gravitational forces are negligible in strength at this separation (although the effect of the weak force can be picked out, feeble as it is, by its tell-tale corkscrew asymmetry even within systems like atomic nuclei that are dominated by the strong force). As the strong force falls off much more rapidly with distance than the electrical force, at a separation of only about 5.6 fm the two are equal (the electrical force in question at this stage is that between the two protons, the circling electrons having negligible effect). At a separation of 1 Å, characteristic of atomic sizes and of atomic separations in molecules, the electrical force is totally dominant, but as the separation increases beyond that the electrical force itself drops off more rapidly than with the inverse square of the distance between the centres of the atoms. This is because, as the two hydrogen atoms separate, the negatively charged electrons come between the positively charged protons and, progressively more rapidly, cancel out or screen the electrical proton–proton force. But the gravitational force, minute as it is, depends only on the masses of the interacting particles and so eventually comes into its own, equalling the electrical force at about 2 mm separation and completely dominating at a few centimetres. Thus although the gravitational force is exceedingly weak (a hydrogen atom held together by gravity alone would be about as big as the entire visible Universe) it utterly dominates in large-scale circumstances of electrical neutrality and holds together the chief structures of the universe – the planets, the stars, the galaxies, the galactic clusters. . . . We are ourselves held onto our Earth by the gravitational force because both we and the Earth are electrically neutral, but if there were suddenly to be brought about an electrical imbalance of only 1 part in 10^6 between protons and electrons, the resultant repulsive force between each one of us and the Earth would be equivalent to about 10^{36} tons weight, which is a measure of the strength of the electrical force relative to the gravitational.

So there appear to be tremendous differences between these four forces of Nature, yet the possibility of there being a fundamental unity between them will be a major theme running through this book and, if there is, this may have dominated events in the first instants of the Universe. But before following up that idea, let us consider a point of philosophical importance: why these four forces? If we take an anthropocentric viewpoint we can see that we 'need' the gravitational force to hold together the stars, to hold the planets in their orbits around the Sun and to hold us onto the Earth; we 'need' the electrical force because it is that which holds the electrons onto their atoms and that holds the atoms together into the molecules out of which we are made; we 'need' the strong force because it is that which holds neutrons and protons together to give the variety of atomic nuclei and hence the range of atomic, and so chemical species essential for building the rich diversity of molecules on which life depends (life could scarcely have evolved from hydrogen alone). But why do we need the weak force? Very simply because it is that which initiates the fuelling of the stars and permits the building of the elements. Stars shine, generate the radiance upon which life depends, because when two protons in the hydrogen of a star bump up against one another, very occasionally one of those protons, in the course of collision, transforms itself through the weak force into a neutron, emitting a positive electron (the antiparticle of the familiar negative electron) and also some energy; the resultant neutron and the other proton stick together as a deuteron, the nucleus of 'heavy hydrogen'. Following this, other, much more rapid nuclear reactions depending not upon the weak but upon the strong and electromagnetic forces lead to the formation of alpha-particles, the nuclei of helium atoms containing two protons and two neutrons, with a considerable release of energy that provides most of the Sun's power. Similarly it is the weak force, with its ability to interconvert neutron and proton through radioactive beta-decay, that permits the building up of heavier atomic nuclei in stars by the addition of neutrons to lighter nuclei, followed, on occasion, by the conversion of neutron to proton inside the nucleus; this increases the positive electrical charge of the nucleus and so it is moved higher up the atomic table of the chemical elements. All four forces are essential if we are to be here to know about them.

There is also another, subtler, condition for our existence which is to do with the balance, within the Universe, of matter and radiation – light of all wavelengths: this balance is about 10^9 to 1 in favour of radiation as measured in terms of numbers of photons, the units of radiant energy, to numbers of neutrons and protons. For reasons to do with the degree of order in the Universe, if this balance were very different we should not be here. Just why these delicate provisions of forces and conditions, critical to our existence, have been made I leave for others to debate.

The tremendous differences between the four forces of Nature have encouraged, rather than discouraged, the quest to find out how, if at all, those forces are interrelated. Are they indeed totally unrelated, each *sui generis*, or can they be linked through some principle of unification so that, in some sense, they are different aspects of each other? Some have believed that they must be related, others that they cannot be related. Einstein spent the last decades of his life unsuccessfully attempting to unify electro-magnetism and gravitation; the former beautifully and probably completely (at least in the classical, pre-quantum-mechanical sense) described by Maxwell's equations; the latter beautifully and possibly also completely (classically) described by Einstein's own equations of general relativity. These unsuccessful attempts were generally regarded as unprofitable exercises; as Wolfgang Pauli remarked: 'What God hath put asunder let no man join'. Nevertheless this attempt to unify electromagnetism and gravity was perhaps natural in view of the inverse square law for the dependence on distance of both the gravitational and electrical forces; but we now understand why the difficulties of that possible eventual unification are greater than those facing the possible unification of the other forces.

Indeed, it now appears, thanks to the work chiefly of Abdus Salam and Steven Weinberg, that the electromagnetic and weak forces have been unified: in 'everyday life' there appear to be two, totally different, forces but they are really just two aspects of a single, more fundamental, interaction. Above a certain régime of energy of interaction between particles, at present unattainable in the laboratory but not totally out of practical reach, the two forces will become of the same strength and will merge into a single force, the electro-weak force, acting equally and indiscriminately between all particles, hadrons and leptons, electrically-charged and electrically-neutral. Hadrons and leptons will still remain distinguished, as before, by the strong force which, at the energy of interaction at which the electric and weak forces merge, still acts only between hadrons and which, with gravity, remains in the wings. The scale of the electro-weak unification energy is about 100 GeV, about 100 times the mass of the proton, so that at energies somewhat about this the two forces fuse fully into the one.

It seems as though the strong force may now, in fact, be emerging from the wings to join the other two. We are in a stage of rich and attractive speculation about what are generically known as Grand Unified Theories (GUTs) that combine the electro-weak and strong forces such that in some yet higher energy régime than 100 GeV the distinction between strong and electro-weak forces is lost, just as above 100 GeV the weak and electromagnetic forces come together. It seems that this GUTs energy régime must be very high indeed: of the order of 10^{15} GeV, 10^{15} times the mass of a proton.

Of the possible ultimate unification of gravity with the other force, in the sense of GUTs, or with the other three forces in the 'everyday' sense, we can only speculate rather blindly although such speculations, the so-called 'super-symmetric' theories, already abound. A chief reason for our difficulty over gravity is that we do not, as yet, have an acceptable quantum theory for it; indeed it is not yet demonstrated that gravity is basically of a quantum nature at all and if it is not it will not be unifiable with the other force or forces without an understanding of a type that we do not yet possess; for the time being the veil is drawn. This is by no means to say that gravity cannot have quantum effects and in fact such are the basis of the mechanism by which Stephen Hawking has described the 'evaporation of black holes'. It is also clear that if we attempt to discuss concentrated masses of greater than a certain magnitude called the Planck mass of about

10^{19} GeV we shall simply not know how to do it because for such masses the quantum *effects* of gravity will dominate the situation, no matter how gravity itself is structured in the quantum sense. For such masses the gravitational force between particles is comparable to the forces of GUTs, the strong and electro-weak forces; thus before such conditions of matter can be considered the problem of the relationship between all the forces, or super-unification, must be understood.

ALPHA AND OMEGA

We are now ready to take a journey in time, back towards the earliest instants of time, if such in any comprehensible sense there were following the 'Big Bang' of instantaneous creation. The basis of most current discussion of the Universe is the Big Bang model which says that in the beginning there was a great concentration of energy, finite in amount but infinite in density, from which the Universe evolved according to immutable Laws of Nature. If this model is wrong and there was not a Big Bang in this sense then ignore everything which follows about the first few minutes of the Universe, but the rest stands. It is also assumed that time and space, the coordinates of the Laws of Nature, are infinitely sub-divisible and do not, themselves, have some sort of ultimate granularity. So far there are no signs of such granularity appearing and, on the contrary, to the finest scales of space and time yet probed (about 10^{-16} cm and 10^{-26} seconds) all seems completely smooth and well. But let it also immediately be said that the times and distances we shall discuss with apparent confidence are vastly beyond this present experimental range of verification and so it could all be vitiated by a granularity not yet perceived. However, our discussion of times longer than 10^{-26} seconds and distances greater than 10^{-16} cm is probably reliable.

At an age of about 10^{-45} seconds the typical energy of constituents of the embryo Universe would have been in the region of 10^{19} GeV, the Planck mass. Earlier than this our primitive understanding, in the absence of the final unification of the four forces, dissolves into an indescribable seething foam of mini black holes, somehow forming, evaporating, and reforming; we are in the realm of total speculation, no longer confident of the distinction of time and space and we must, at least until the great steps to final unification are taken, be silent. Even after the Planck time has elapsed the Universe is not all plain sailing from the point of view of our present physical understanding. In particular, we must understand why our Universe today consists almost solely of particles, with very few antiparticles. In the beginning, unless the Creator wished to impose special conditions upon His creation, there was simply energy: a finite amount in all probability, but infinitely concentrated. Now when, in today's laboratories, we simply let energy loose through the violent collision of accelerated protons, for example, new particles and their antiparticles are created together out of that energy, following Einstein's $E = mc^2$, in equal abundance. Why, then, is the Universe itself, also, as we rather confidently speculate, made originally just out of energy, so lop-sided in the particle versus antiparticle sense? We must assume that at the end of the Planck time, after the eventual lapse of those 10^{-45} seconds, the evaporation of all, or most of, the black holes gave us, in a way that we do not yet know how to discuss, the beginnings of our modern Universe with equal numbers of particles and antiparticles. However these were not the particles and antiparticles found in today's Universe, neutrons, protons, electrons, etc., but rather progenitors of those particles: progenitor particles whose own radioactive decay eventually gave rise to the building materials of the edifices – including ourselves – of our present interest. These are new ideas emerging from the attempts to construct GUTs. The progenitor particles entered the scene after gravity slackened its grasp with masses at the GUTs unification energy of about 10^{15} GeV. . . . starting with equal numbers of progenitor particles and antiparticles, the present preponderance of one sort of matter over the other came into being; a catastrophic self cancelling annihilation of equal amounts of matter and antimatter, creating a universal flood of radiation, was averted, leaving the observed, 'needed', ratio of about 10^9:1 between radiation and matter by the time the Universe was 10^{-35} seconds old.

All this does not, of course, prove that GUTs are correct, but it does suggest that we are not necessarily begging any fundamental questions in dating the modern history of the evolution of our Universe from about 10^{-35} seconds after Creation, complete with the implied preponderance of particles over antiparticles and with the implied balance of radiation as against matter.

After the first 10^{-35} seconds we believe things took a relatively conventional course based on physics in which we can feel greater confidence, better grounded in observational fact, many unknowns and uncertainties as there still may be. One major uncertainty concerns the very internal structures of the neutron and proton themselves and of all other hadrons; a related uncertainty is the nature of the strong force. So far we have spoken of the strong force as though, in its essence, it acts *between* hadrons; however, the basic strong force is more likely to be that acting *within* hadrons, that *between* hadrons being no more than an external reflection of this. The direct descendants of the progenitor particles are thought to be not the neutrons, protons, and the other hadrons themselves but rather the *constituents* of those particles, the quarks, between which the true basic strong force has its play; but more of this later.

The GUTs may offer an explanation for the origin of the matter we are made of but it seems difficult to take seriously a theory of events occurring at a time of zero plus 10^{-35} seconds we can scarcely imagine and at an energy of 10^{15} GeV that we can never achieve in our laboratories. (An accelerator wrapped around the entire equator of the earth and equipped with the most powerful magnetic field we know how to make to guide its particles would generate an energy of only about 10^7 GeV; such an accelerator for 10^{15} GeV would be a hundred times bigger than the Solar System.) But there is a point which opens the theory to practical enquiry: the GUTs make a dramatic, unexpected, and testable prediction, namely, that the proton itself must be unstable and eventually decay radioactively. The very forces that brought us into being also ensure our eventual destruction. There is, however, no cause for immediate alarm because the lifetime of the proton predicted by GUTs is about 10^{32} years, almost infinitely large when compared with the present age of the Universe, about 10^{10} years. The current experimental lower limit on the proton's lifetime is already about 10^{30} years, however the GUTs prediction is accessible to test and although the experiment is very difficult several searches are in preparation. Paradoxically, an experiment telling us something about a time scale of 10^{32} years will give us confidence in talking about events on a time scale of 10^{-35} seconds. For those concerned about a life-span of only 10^{32} years I must add that more elaborate GUTs can, in fact, be constructed in which the proton is stable and that there are others who believe that even with decaying protons the rate of that decay decreases as the Universe ages so that only a minute fraction of the total would ever decay.

So by about 10^{-35} seconds we had the progenitor particles, the nature and variety of which depend on your choice among the GUTs, in the mass-energy range of 10^{15} GeV. The radioactive decay of the progenitor particles gave rise to the internal constituent particles of the hadrons, the quarks. As the Universe expanded particle–antiparticle annihilation took place, but a minute excess of particles survived, and eventually the density was sufficiently low for the quarks to cluster into the familiar hadrons: neutrons and protons and other particles that we know within the laboratory. This happened after about 10^{-6} seconds. As time passed the more highly excited of the product hadrons quickly disappeared and after about 10^{-5} seconds only neutrons and protons and a few of the lightest of the other hadrons, together with leptons of various kinds and, of course, photons were left. Beyond this time the evolution of the Universe depended chiefly on the number of distinct species of lepton that exist and related matters. Conventional nuclear and particle physics, accessible to laboratory investigation, had already taken over, the time scale stretches enormously and after about 3 minutes the course was set; most of the Universe's helium together with some other light nuclear species had already been manufactured; the rest is history.

But where are we today? What is the Universe like now and what will happen to it in the future? The Universe is made of stars, largely clustered into galaxies of various kinds, many of the galaxies flat, disc-like, and roughly spiral-shaped such as our own Milky Way; then there are odd things such as quasars that emit tremendous amounts of energy in a manner as yet by no means understood; probably some, perhaps many, black holes, a lot of gas, a lot of dust, and so on. But the whole thing is expanding as though everything were still rushing apart following that Big Bang. Will it go on expanding forever (although perhaps rather radically transformed after 10^{33} years or so by the GUTs decay of the protons and neutrons)? It is not certain that the Universe will expand forever, because gravity is obviously acting to slow down the expansion, to stop and reverse the motions, and bring everything back to where it started from; a Big Crunch followed perhaps by another Big Bang, or a Big Bounce. It is a question of how much mass there is in the Universe; if there is more than a certain mass then gravity will win in the end and everything will collapse into a Big Crunch – the Universe, in the jargon, is closed; if there is less than this mass then the expansion will continue forever – the Universe is open. Bookkeeping on the mass is difficult because of the amounts of gas and dust that lie, in quantities difficult to determine with precision, within and between the galaxies, because of the unknown number and mass of black holes, and so on. Indeed, there are within astronomy, at one or two places, what the astronomers call 'missing mass' problems. If you look at the motions of stars within galaxies you should be able to explain the dynamics in terms of the mass of the galaxy and the distribution of that mass within it, but when this is tried it does not come out right: it seems that there must be more mass within the galaxy than has so far been detected. Similarly when you look at clusters of galaxies and analyse the motions of the galaxies within the clusters, mass again seems to be missing. But even taking these problems into account present best estimates suggest strongly that there is not nearly enough mass there to close the Universe.

The open or closed Universe is not as yet, however, an open and closed book; there is a major point of reservation to be made. I have spoken of the weak force as transforming, for example, a neutron into a proton plus an electron, but this is not all; a third particle is emitted in that radioactive decay, an electrically-neutral lepton, the neutrino. The neutrino, whose interaction with both electrons and hadrons has been investigated experimentally in great detail, may be massless and has long been popularly supposed to be so. Now enormous numbers of neutrinos must abound in the Universe (the scenario we have followed implies that they are roughly equally as numerous as photons and so about 10^9 times as numerous as neutrons and protons). If neutrinos are massless they make no contribution to the mass of the Universe and are irrelevant to the question of the open or closed nature of the Universe. But they may have a mass and, indeed, very recent experiments suggest that the mass of the neutrino accompanying the beta-decay of the neutron is finite and perhaps as great as a few thousandths of a per cent of that of the electron. In that case the total mass of the neutrinos in the Universe would be so great as vastly to outweigh (literally) all other known matter, and could be sufficient to close the Universe, which would then collapse back into the Big Crunch in perhaps another 10^{10} years or so. No sooner are we relieved to learn that GUTs proton decay nevertheless gives us another 10^{33} years than we find that we may be cut off after only 10^{10} years!

Before leaving the fate of the Universe it should be remarked that Einstein's equations of general relativity, on which the gross time-evolution of the Universe is thought to depend, admit of the addition of a term, the so-called cosmological constant, which could have the effect of forcing an expansion on the Universe even if the mass of the Universe were such as would, in the absence of that constant, close it. This cosmological constant was originally introduced when it was mistakenly thought to be necessary to secure acceptable solutions to Einstein's equations, but when the mistake was discovered it was seen to be unnecessary and so fell under the slash of Ockham's razor. But it could still be there and could, despite a finite mass for the neutrino, save us from the Big Crunch.

THE UNIVERSE NOW

So how big, irrespective of its ultimate fate, is the Universe now? One can feel one's way out to this, starting from things with which one is familiar. First of all let us try to get a feeling for the size of the Earth. The curvature of the Earth is appreciable, although not usually perceived over short distances; if the ground were smooth and free of obstructions, if you went out of Wolfson College and, with your eye at ground level, looked down the road the 1.5 miles to Christ Church, the curvature of the Earth would cause the Dean to appear to be buried to his knees. You have a good grasp of 1.5 miles from ordinary personal experience, you have a good grasp of one-third of a Dean, you can begin to *feel* the curvature of the Earth. We could extend this picture to visualize the top of Salisbury cathedral's spire, 50 miles away and 1200 feet below the horizon; and so on. But today we are used to a view from the Moon: at a distance of 240 000 miles we see the Earth as a small globe, 8000 miles in diameter. Nearly 400 times further from the Earth, at 93 million miles, we find the Sun; this is a big step and now we should change to a more appropriate measure for big distances: the time taken for light to cross them. At 186 000 miles a second it takes Moonlight a little over one second to reach us, and Sunlight about 8 minutes. The nearest star to the Sun is α-Centauri which is about 4 years away, but there is another way of getting a feel for that if you have a feel for the distance of the Earth from the Sun and therefore the diameter of the orbit on which the Earth circles the Sun. Imagine, as the Earth goes around the Sun, that you fix your eyes on α-Centauri: because the Earth is moving around the Sun, α-Centauri will seem to move to and fro relative to the more distant stars that form the back-drop. The apparent motion of α-Centauri is about a second of arc and that is about the angle of a thumbnail at 3 miles, which you can picture all right. Beyond the nearest star distances begin to get quite appreciable and the effort to seize them also gets quite considerable. Our galaxy, the Milky Way, containing a few thousand million stars, is about 100 000 years across (although much thinner, being a spiral pancake); our nearest sister galaxy, Andromeda, is about a million years away but that is only a few hundred thousand times the distance of α-Centauri that you pictured through your thumbnail. Beyond Andromeda we had better go at once to the remotest depths of observable space, only about ten thousand times as far as Andromeda, and that is the end as far as we can see. The Universe is huge, without doubt, but you can almost grasp it, provided you can imagine the Dean of Christ Church sunk to his knees.

Nor is the observable Universe all that big in relation to the things it contains. The largest apparently structured objects that we know are observed not in visible light but at radio wavelengths through the techniques of radio-astronomy; they look rather like 'bow ties' and their appearance is due to electromagnetic rather than to gravitational forces. These objects are ordered streams of electrically charged particles spanning distances of some tens of millions of light years and so they stand to the entire Universe, as we can see it, just about as does a bow tie on the neck of a lecturer to a large lecture theatre. Large as the Universe may be, the forces that shape it manifestly reach out over distances not incommensurable with the totality of its size: imagine a lecture theatre with a large number of bow ties within it – we would scarcely say that it was empty.

ATOMS

I have tried to persuade you to accompany me to the edge of the Universe, to persuade you at each stage that your imagination could take the next step. Let us now reach inwards towards and into the atom, to scales of distance vastly less than our own personal dimensions just as those of the Universe are vastly greater. Indeed, in discussing the earliest instants after the Creation we have already assumed there is sense in the discussion of objects that are smaller than ourselves by factors much larger than that by which the visible Universe is greater. What right have we to suppose that in either of these realms, so much larger and so much smaller than that of which we have direct personal experience, the language that we have devised to describe our personal experiences of events more or less of our own scale will continue to work? Lucretius had the same problem:

> The poverty of our language and the novelty of the theme compel me often to coin new words. This it is that leads me to stay awake through the quiet of the night studying how by choice of words I can display before your mind a clear light by which you can gaze into the heart of hidden things.

Similarly, how can we be sure that the mathematics that we have devised to codify our everyday experience will continue to work under conditions remote from that experience? May it not be that on scales of time and distance vastly beyond the realms of our personal experience, phenomena will obtain that are simply not susceptible of description through any thought process of which we are capable? That is to say, in our human terms, may there not be essential irrationalities in the very big and the very small such that we can never grasp their workings because to us they could never 'make sense'? This is an open question and I would beg you to leave it open, at the same time recognizing that the name of the game is to try to describe what we cannot possibly reach as though it obeyed the same rules as what we can. Recognize, as a corollary, that we may not be describing and understanding the Universe, very big and very small, but rather inventing it.

So now down to the atom. It is not very far away. We can easily see, through an ordinary microscope, things as small as 10^{-4} cm across – about a millionth of our own size. The atom is only about ten thousand times smaller and nowadays we can see that as well, not by bouncing light off it as one sees something with one's own eye but by bouncing high-energy electrons off it and then focusing them through electric lenses much as light is focused through glass lenses. What we then see (with the naked eye) on a TV-type screen exposed to focused electrons that have been scattered off a thin slice of crystal is a regular array of dots, each due to the scattering of the electrons from an ordered line of atoms in the crystal just as chemists and X-ray crystallographers have, for decades, been telling us we must. There are no surprises but it is nice to see the atoms properly regimented in their ordained rows. We can similarly 'see' single isolated atoms of heavy elements such as uranium scattered randomly over some light supporting surface and, more impressively, we can 'see' single molecules that contain more than one heavy atom, the separation between the heavy atoms tallying exactly with our earlier inferential understanding. With electron beams we can also, by more indirect means analogous to the now-familiar optical holograms, go inside individual atoms and map out the distribution of their electons in the 'line of sight' at various distances from the centre of the atom, as if you had plunged a minute cork borer through the atom and counted how many electrons you had bored out with it. Here also the results confirm what we had known for a long time, including such superficially surprising things as that the light atom neon has more electrons in certain regions of its anatomy than the heavier atom argon.

Atoms are like little Solar Systems somewhat more than 10^{-8} cm (1 Å) across, with electrons circling the central nucleus, itself only about 10^{-13} cm in diameter, made of neutrons and protons; the number of protons is the atomic number and determines the chemistry of the atom; the nucleus carries all but a few thousandths of a per cent of the mass of the atom. We should not say that the electrons 'circle' the nucleus, because quantum mechanics, the ground rules for talking about very small objects, does not permit the unqualified use of household concepts for discussing the very small, although what it requires us to use in their place are rather mixtures of household concepts; it does not require us to abandon everyday speech in our descriptions and certainly does not force us towards the irrationalities about which I speculated on switching from the very big towards the very small. In place of 'circling' as a planet does the Sun, the electron simply has its being within the atom, has a certain precisely defined probability, or chance, but only a *probability*, of being found at any spot if we suddenly dive into the atom at any arbitrary time. The electron also has associated with it a certain, also precisely defined, angular momentum of its motion around the nucleus of the atom. Quantum mechanics forbids us actually to watch the electron going round and round, or rather tells us that if we try to watch it we shall so gravely disturb its motion that the results of the observation will not be interpretable in terms of what would have been happening had we not looked; so there is no point in talking about 'what would have been happening' at all, since we can never check on it. Of course, planets going round the Sun should also be described quantum mechanically because quantum mechanics is a more fundamental law than classical mechanics, but we observe planets perfectly happily and do not fear knocking them off their orbits by so doing. There is no contradiction here: quantum mechanics tells us that the bigger a system becomes the more nearly do the laws of quantum mechanics approximate to those of classical mechanics and, indeed, even for an atom it becomes permissible to talk of an electron as circling its central nucleus if the atom if sufficiently 'excited' to move the electron out into an orbit of sufficiently large angular momentum.

How do we know that atoms are little Solar Systems (using that term within the above quantum-mechanical stricture)? Apart from 'seeing' them with the electron microscope, we calculate their properties, using quantum mechanics, and then carry out measurements to check those properties. In particular we can measure how much energy we have to put into an atom to remove an electron – that electron's separation energy – and compare that with calculation. When this is done we find very striking regularities: as we go to heavier and heavier atoms the amount of energy needed to prise loose an electron steadily increases, as we might expect on account of the increasing electrical attraction of the increasing positive charge on the central nucleus, but every now and again there is a sudden drop in the energy needed, followed by a subsequent rise from the new low. This is precisely what is predicted by the quasi-Solar-System model coupled with an extraordinary quantum mechanical rule called the Pauli exclusion principle, which limits, rigorously, the number of electrons permitted to occupy one 'planetary orbit' and demands that the next go into a more remote, and therefore less strongly bound, orbit.

So we have very good confidence when we speak of atoms and of their building up into molecules of varying degrees of complexity, from the simple H_2O to the giants that are associated with the processes of life. However the statement that an atom is only ten thousand times smaller than something that you can actually see with an ordinary microscope may perhaps give the impression that atoms are large, ungainly galumphing things; so they are, compared with what we shall look at next, but they are, nevertheless, quite small and very numerous. Classical illustrations of the abundance of atoms and molecules have always been popular: that your next cup of tea will contain about a thousand molecules of water from the cup of hemlock that Socrates was forced to drink, assuming that Socrates' cup has by now been fully mixed into the oceans and atmosphere of the world; that, at this moment, the lungs of each one of you contain about one molecule from the expiring breath of Julius Caesar (and, for that matter, of Genghis Khan and Helen of Troy . . .), again assuming a uniform mixing of expiring breaths throughout the atmosphere. But there is no need to over-sensationalize a theme, which is surely dramatic enough in itself.

Now to the central nucleus, the Sun of the atom. How do we know that it is there and how big is it? If one fires accelerated atoms into a gas, or a thin foil of matter, they will be deflected from a straight path, or scattered, because they contain electrical charges and so do the atoms on which they impinge. (Although individual atoms are electrically neutral, electrons can be got out of them, as we have just noted, and so the overall electrical neutrality must be made up of some sort of balance between negative and positive constituent charges.) We can measure the probability that the impinging atom is scattered through very small angles and then can confidently predict the probability that the impinging atom will be scattered through a large angle *if* that is just due to the chance accumulation, as it passes through or near many atoms of the target gas or foil, of many small nudges in the same direction. The result is sensational: the probability of scattering through a large angle is totally inconsistent with expectation derived from the scattering through small angles. In the experiment carried out by Rutherford in 1911 swift helium atoms impinged on a thin gold foil and the probability of scattering through more than 90° was about $10^{10\ 000}$ times greater than expected from scattering through small angles. It is impossible to imagine so large a number; it is much easier to hold the Universe in the palm of your hand. Rutherford said that he was as astonished as if a 16-inch shell had bounced backwards from a sheet of tissue paper. Obviously atoms are something other than just vague balls of mixed positive and negative electricity. Rutherford's quasi-Solar-System model to explain this tremendous large-angle scattering was that at the centre of each atom is a 'point-like' nucleus carrying as much positive charge as all the circling electrons carry negative charge and also carrying almost all the mass of the atom. When, now, two atoms impinged 'head on' the two nuclei approached each other and, with the entire positive charge within each atom being concentrated at a point rather than spread out through the entire volume of the atom, the nuclei could exert vastly greater forces upon each other, as they came close together, than if the charges had been spread out. The spread-out model was a good enough approximation to describe the small-angle scattering when the nuclei remained far apart, but not when they came into close collision. It is easy to work out the law of scattering probability as a function of angle of scattering for this nuclear model and it fitted perfectly: the discrepancy of $10^{10\ 000}$ disappeared.

Speaking of the positive charge of the nucleus we used the phrase 'concentrated at a point'; this, of course, is not so but it works as though it were true so long as the nuclei do not touch. If the nuclei do touch then the simple point-scattering law breaks down and from the condition under which this happens we can infer the size of the nucleus: about the 10^{-12} cm mentioned earlier. Since, so far as we know, electrons are point charges and since the atom is about 10^{-8} cm in its own dimensions the volume occupied by the nucleus is only about 10^{-12} of that of the atom: the atom is almost entirely empty space.

I have several times mentioned neutrons and protons: they are the constituent particles of the nucleus, almost identical to each other in mass and other properties apart from their electrical charge which for the proton is plus one (the electron being minus one) and for the neutron zero. The neutron and proton themselves have size – about 10^{-13} cm – as we can discover by bouncing them off each other or by bouncing electrons off them. When the electron penetrates inside a proton it feels a smaller attraction than if the proton had been a point because part of the proton's charge is then trying to pull the electron out and only the remainder to pull it in, so the resultant scattering is through a smaller angle than it would have been for a point proton; in this way the proton's (electrical) size can be measured and, indeed, the distribution of its charge effectively mapped. The neutron, although electrically neutral over-all, also scatters electrons because it has an internal electrical structure that, as for the proton, can be mapped by the scattering.

The neutrons and protons inside the nucleus are not packed tightly together and indeed they move quite freely around in almost independent orbits like the electrons far away in the outer space of the atom's Solar System. We have, in effect, a quasi-atomic model for the nucleus itself, the nuclear shell model; there is no special centre to the nucleus, it has a uniform density and the binding force experienced by any neutron or proton is just the sum of the forces on it due to all the other neutrons and protons of the nucleus. The shell model shows up phenomenologically in much the same way that the electronic orbital model of the atom did: by systematic changes, with increasing mass of the nucleus, of the neutron and proton separation energies and with sharp drops in those energies when the next neutron or proton is forced, by the Pauli exclusion principle, into a higher orbit.

There is a tremendous wealth of nuclear phenomena; we know and have studied the properties of many hundreds of nuclear species, each being characterized by its own number of neutrons and protons. By-and-large we understand what is going on in terms of many collective phenomena which are all entirely consistent with the quasi-atomic shell-model and which, indeed, can be derived from it but which form more succinct and appealing ways of describing certain of the modes of organization of the neutrons and protons: we speak of the nucleus as being deformed into ellipsoidal or hour-glass shapes, of rotating or of vibrating, with all the possible combinations of these gymnastics; we speak of oscillations in which all the protons and all the neutrons swing backwards and forwards relative to each other in two inter-penetrating sets; and so on, but none of these ways of speaking is other than a convenient short-hand to summarize what would be a somewhat elaborate description in terms of the quasi-atomic shell model.

Here we shall leave the rich variety of nuclear phenomena; they still pose many interesting questions but in the present context of an attempt to paint, with an exceedingly broad brush, the whole range of our vision of the organization of the Universe, we can consider nuclei to be fairly well understood.

MESSENGERS OF FORCE

Before moving to sub-nuclear matters we must go right back to the beginning when I spoke of the four forces of Nature. How do forces come about? What makes the inverse square law of the electric force and the more sharply cut-off law of the strong force, which drops very quickly to zero after about 10^{-13} cm? There is, among physicists, an abhorrence of 'action at a distance'. Crudely speaking we have a revulsion from any theory that speaks of an interaction – a force – between particle A and particle B without saying how particle A becomes aware of particle B's presence. In other words we demand (admittedly on philosophical, or possibly even sentimental, grounds) that interaction should depend on communication: A and B cannot know, cannot experience a mutual force, attractive or repulsive, unless they find out, via an appropriate messenger, about each other's existence. The messenger, of course, can only be some other particle which A emits and tosses across to B and vice versa.

We therefore believe that all forces in Nature (pace, for the moment, gravity which remains somewhat enigmatic) are generated by the exchange, between the interacting particles, of messenger particles. Let me give a homely illustration of a force generated by a messenger particle. You are observing, from afar, two skaters on a frozen lake: they are drifting along side by side, a few feet apart, but then, without apparent movement of their legs, begin to move further apart: a repulsive force appears to exist between them. You put up your binoculars and observe that although they are doing nothing else they are throwing a cricket ball to and fro. Skater A throws the ball to skater B; A recoils from his throw, moving away from B and B recoils from his catch, moving further away from A; and so on and so on. The cricket ball is the messenger particle that has established a repulsive force between A and B. You may accept this home-spun illustration of the establishment of a *repulsive* force due to exchange, but how about an attractive force? Consider the same example of skaters on a lake but imagine that, from afar, you see them coming closer together: the force between them is now attractive. Put up the binoculars again: this time they are exchanging a *boomerang*. Skater A throws the boomerang away from skater B and A therefore recoils *towards* B. The boomerang curves through the air away from him and then towards B, passes him and then curves back from the other side so that B catches it when it is moving towards A and therefore gives B a push *towards* A, the force is attractive; and so on and so on. (There will be those among my readers who will consider, and quite rightly, that this illustration is skating upon thin ice.)

What are these messenger particles? In the case of the electromagnetic force they are the photons, the quanta of electromagnetic energy. The nuclear force can be communicated by a whole host of messenger particles called mesons. Now the reach of a force, the extent of the space over which its influence can be felt, depends on the mass of the messenger: the heavier it is the shorter the range. Thus the photon which has zero mass (although it does carry energy) may, technically speaking, extend the influence of the electromagnetic force to infinity. (The famous inverse square law arises simply because the area of a sphere of radius r increases as r^2 so that the chance that a photon despatched randomly by A will light upon B at a distance r goes as $1/r^2$.) The lightest meson is the pion with a mass about 275 times that of the electron and this explains why the strong, nuclear force drops off so sharply beyond about 10^{-13} cm. There are many other mesons of greater mass that contribute to the strong force at smaller distances and which give that force a rather complicated form. The weak force we believe to be transmitted by particles, also mesons, called intermediate vector bosons, $W^{\pm}$ and Z^0, of mass about 100 times that of the proton and hence limiting the reach of the force to a few times 10^{-16} cm. Thus we picture the beta-decay of a neutron into a proton plus an electron plus an antineutrino as proceeding by the emission of a W^- particle from the neutron, which latter thereby becomes a proton, followed 10^{-26} seconds or so later by the disintegration of the W^- into the electron and the antineutrino. These $W^{\pm}$ and Z^0 particles are intimately linked to the photon and specify the energy scale relevant for the electro-weak unification of the electromagnetic and weak interactions. The possible GUTs unification energy of the electro-weak and the strong interactions is also associated with a messenger particle, named *lepto-quark*, whose mass would be of the order of the unification energy of about 10^{15} GeV; this same particle is the progenitor, the one with such a crucial role in the birth of matter near the beginning of this Universe.

PROTON STRUCTURE

We now turn to the proton and neutron themselves. They used to be called 'elementary particles' but that name is clearly no longer appropriate if by it you mean a thing entire unto itself, having no constituent parts or wheels going round inside it. That protons, in fact, have a structure can be inferred in several ways. Perhaps the most appealing is the direct analogue of Rutherford's experiment on the scattering of swift atoms by matter that revealed the nuclear structure of the atom so dramatically, showing unequivocally that atoms have little hard bits inside them. When protons bombard each other at very high energies they tend to emit mesons, as is only to be expected since it is by the emission and absorption of mesons that protons communicate with each other. If protons are simply pieces of matter of some sort, with extension but no internal structure, then, just as in the case of the highly-probable small-angle scattering of Rutherford's experiment, we can observe the copious emission of mesons at small angles relative to the collision axis of the protons, emission corresponding to the relatively gently nudging of the protons by each other, and confidently extrapolate the production to large angles of meson emission. More accurately, we extrapolate the production of mesons with a small transfer of transverse, sideways, momentum between the participants to the production of mesons of high transverse momentum transfer. When we do this we get the same shock as Rutherford: at high momentum transfer the production of mesons is many orders of magnitude greater than we expect from our extrapolation of the low momentum transfer production. Just as surely as there are little hard bits inside atoms there are little hard bits inside protons to give that unexpectedly frequent high transverse momentum transfer; protons and neurons and, we presume, all hadrons are structured objects containing constituent particles much smaller than themselves.

The best limits on the size of these constituent particles come from bombarding protons with very energetic neutronis, point-like particles interacting only via the weak force. Let us assume, for the moment, that the weak interaction between the neutrinos and any proton constituents is of effectively zero range. Then, for reasons to do with basic quantum mechanics, if the constituents within the proton can be regarded as point particles the probability that a neutrino, on impinging upon a proton, causes it to emit another particle – a lepton of some sort – will be proportional to the energy of the neutrino. But if the constituent sub-structure of the proton, with which the neutrino is interacting, itself has finite size then this simple proportionality will break down; the energy of neutrino at which this happens is a measure of the size of the constituent particle. So far, in experiments involving neutrinos of energies up to 100 GeV, this proportionality holds well enough for us to state both that there are 'little hard bits' inside the proton no bigger than, at most, 10^{-15} cm and that any finite range of the weak force is also less than this.

(Indeed, the predictions of the electro-weak theory suggest the range is about 10^{-16} cm.) Since the proton itself is about 10^{-13} cm across this small limit on the constituent size suggests that protons and neutrons themselves, like the atom, are chiefly empty space. . . . [These experiments with neutrinos] enable us to count the number of 'little hard bits' in the proton: it comes out at 3 or close to it.

Another, totally different, line of argument also suggests strongly to us that the protons, and all hadrons, are indeed made out of some sort of constituent particles. If we take any structured system: a molecule made out of constituent atoms, an atom made out of constituent nucleus and electrons, a nucleus made out of constituent neutrons and protons, we can rearrange the constituents of that system relative to each other and thereby raise that system into states of higher energy, called excited states. We can make the atoms in the molecule vibrate against each other, we can elevate the electrons of the atom into higher orbits, we can make the neutrons and protons of the nucleus vibrate or move to higher orbits; all structured systems exhibit a tremendous richness of excited states simply because their constituent sub-units can move relative to one another in such a variety of ways. Indeed, if one is shown a diagram . . . of the excited states of one of these objects, molecule, atom, or nucleus, one cannot tell to which object the diagram refers without being told the energy scale: meV or eV for molecules, eV for atoms, keV or MeV for nuclei. Now if energy is given to a proton, say by bombarding it with another proton or by any other means, it can be raised into a whole host of excited states, just as can the other structured systems, the only difference being the energy scale: not now meV or eV or keV or MeV but rather tens or hundreds of MeV. The same is true of the mesons; they too can evidently be raised into excited states. There is no avoiding the conclusion that the proton, and all other hadrons are, like the nucleus, the atom, and the molecule, structured objects.

What are these constituent particles, the 'little hard bits' of the proton and all hadrons? To cut a long and intensely fascinating story very short: when one looks at the systematic relationships between the excited states of all the hadronic particles that are now known – the number of excited states running into hundreds thanks to enormously painstaking work over the past 30 years at the great accelerators in the USA, Europe, and the USSR – we find that it all makes sense in terms of the constituent particles that Murray Gell-Mann dubbed *quarks*. Hadrons such as the proton should contain 3 quarks (recall the experimental figure quoted above from the neutrino experiments) while the mesons should contain one quark and one antiquark. We know nothing to contradict this beautifully simple picture; the excited states are simply rearrangements of the constituent quarks relative to each other like the rearrangements of the electrons that give rise to the excited states of their atom, or the atoms whose rearrangement gives rise to the excited states of their molecule, or the neutrons and protons whose rearrangement gives rise to the excited states of their nucleus.

There are several species or 'flavour' of quark; there are 'ordinary' quarks (of two kinds called 'up' and 'down') such as go into the neutron and proton and their excited states; there are 'strange' quarks that, by themselves or mixed with the 'ordinary' quarks, made the 'strange' particles that come out of energetic collisions at the great accelerators but that take so long over their subsequent radioactive decay that it is as astonishing as if Cleopatra had fallen off her barge on the Nile in 35 B.C. but splashed into the water only today; there are 'charmed' quarks that have other distinctive properties again and 'bottom' quarks and, probably, but not yet pinned down, 'top' quarks. And there, perhaps and perhaps not, the list will close. But all the myriads of hadronic particles that we know of can be constructed out of this short list of quarks and their antiquarks.

A very obvious question: what holds the quarks together inside the proton and the other hadrons? An associated question is why we have so far failed, despite the blandishments and importunings of our most powerful accelerators, to liberate individual quarks from hadrons no matter how energetically we have bashed hadrons together. Atoms can be got out of the molecules that they compose; electrons, part of the structure of atoms, can be liberated from atoms; nuclei can be dissociated into their constituent neutrons and protons; why, if quarks are inside neutrons and protons cannot they be got out of them? Lucretius would not have worried about this:

> There is an ultimate point in visible objects which represents the smallest thing that can be seen. So also there must be an ultimate point in objects that lie below the limit of perception by our senses. This point is without parts and is the smallest thing that can exist. It never has been and never will be able to exist by itself, but only as one primary part of something else. Since they cannot exist by themselves, they must needs stick together in a mass from which they cannot by any means be prised loose.

But today we must confess that we must find an answer to why we cannot prise loose something that we confidently assert to be inside. (Do not confuse this with the inverse problem: the fact that something comes out of something does not require that it should have been there in the first place; an electron and an antineutrino come out of a neutron leaving behind a proton, as we have noted, but they were not there to begin with; both were manufactured in the act of radioactive beta-decay of the neutron into a proton; barks come out of dogs but dogs are not made of barks. However, if we say that something *exists* inside something we have to be able to say why we cannot get it out.)

On the guiding principle that force is communicated by the exchange of a messenger particle between the interacting objects, the answer to the first question, namely what holds quarks together inside hadrons, is that *within* hadrons, acting between the quarks, is a particle, somewhat unimaginatively named the gluon, that sticks quarks together to form hadrons. The second question as to why we do not seem to be able to prise quarks out of hadrons is not yet satisfactorily answered but a partial answer is that when you look into the force that is generated between quarks by gluon exchange you find that, unlike the other forces we have encountered, it *increases* as the separation between the particles increases, rather than decreases; the closer the quarks, the *weaker* their interaction. (Surprising but not perhaps all that surprising – think of a rubber band.) This property, technically known as quark confinement, is associated with another property of quarks. It seems that each flavour of quark – up, down, strange, charmed, bottom, and (we anticipate) top – comes in three varieties called the 'colours'. Thus we speak of, say, red, yellow, and blue 'up' quarks and similarly (the same colour range) for the others. (It must surely go without saying that, in this context, colour is only a label, a name, and has nothing to do with the red, yellow, and blue of our own perceptions; we could use any other name and refer, for example, to Conservative quarks, Labour quarks, and Liberal quarks.) The reasons for this extra label of colour are compelling but rather technical. . . . It turns out that the gluons that flit between the coloured quarks must also carry colour and from this follows naturally, but by no means obviously, the property of confinement or at least an interaction that increases with distance of quark–quark separation until we meet the analogue of the snapping of a rubber band. But even this does not release the quarks; when the energy put into the system through the effort of separating the quarks against the force of confinement exceeds the energy needed to make mesons then one, or more, of these particles is created; we get new quark–antiquark pairs but no liberated quarks.

This quark–quark force is the strongest of any of which we are yet aware; the strongest force we know in Nature is that which operates not *between* neutrons and protons but *within* neutrons and protons. How is this related to what we earlier termed the 'strong' force, that which operates *between* neutrons and protons, holding them together inside the atomic nucleus? There is more than one way of answering this question but the best present answer is by analogy with the force that operates between atoms as they approach each other. Within each individual atom we have the inverse square law of electric force between the nucleus and the electrons and between electron and electron, which, following the rules of the quantum mechanical game, establishes the structure of the atom. But when one atom looks at another from a distance it is not, as we have already noted, electrically impressed, because the viewed atom appears to the viewer atom as electrically neutral. The summed electron charges of the 'orbiting' electrons are exactly balanced by the positive charges on the protons of the central nucleus; although the electric force is an inverse square force with distance it has totally cancelled itself out because of the over-all electrical neutrality of the whole atom. Indeed at sufficiently large distances of separation the extremely feeble gravitational force then wins out over the electrical force. But now, as the two atoms tentatively approach each other they begin slightly to interpenetrate; electrons 'belonging' to one atom may, if they stray far enough away from their parent nucleus, sense the presence of the other atom's electrons and nucleus. As they interpenetrate a force is established between them, exceedingly feeble at first but increasing very rapidly as the interpenetration increases. This atom–atom force is known as the 'van der Waals force' and, empirically, looks very different from the inverse square electrostatic force; it increases, indeed, something like the inverse seventh power of the distance between the atoms. However it is in no fundamental sense a force in its own right but only a representation of a particular aspect of the inverse square law of electrostatic force operating within the laws of quantum mechanics between extended and structured electrostatic systems that each possess over-all electrical neutrality. In the parallel situation of the quark–quark force due to gluons and the proton–proton force, protons, remote from each other, have no strong interaction to speak of (their electrical interaction is irrelevant here: if you wish, think of the interaction between two neutrons) because the gluons are fully occupied with the intra-proton forces that hold the proton's constituent quarks together. But as the protons approach each other, the quarks in the one proton sense the presence of the quarks in the other. How do they do this? Is it by the exchange, between the quarks of the one proton and the quarks of the other, of the same gluons that give rise to the quark–quark force within each proton? This is tricky to arrange in terms of the book-keeping of colour but it might be done. Is it by the actual percolation of the quarks themselves between the protons? Is it by the creation, through the gluon force, within one proton, of a fresh quark–antiquark pair which then, as a 'conventional' meson, trips to the other proton? The answer is probably complex: at large distances of separation the last possibility is probably right and at short distances the other two mechanisms must certainly come into play. But however we look at it, the strong force between two protons is due to a sort of leakage, from one proton to the other, of the basic strong gluon-exchange force that operates *inside* each proton between its constituent quarks. In this, qualitatively clear sense the proton–proton strong force stands to the fundamental quark–quark strong force much as the atom–atom van der

Waals force stands to the basic electrostatic force within each atom.

EPILOGUE

That is about as far, but not quite as far, as I should wish to go in this stage-setting introductory chapter. There is one outward-looking and one inward-looking final remark that I should wish to make. The outward-looking remark: What is so special about us and our Universe? We do not know that there are not many more Universes, perhaps constituted of different sorts of matter and containing different sorts of consciousness from our own, perhaps even – most excitingly – capable of exchanging exceedingly weak but detectable signals with our own; the anthropic principle, that says that our Universe has to be as it is or we should not be here to know that it is as it is, is a powerful one but also tells us that there are indeed other Universes elsewhere and elsewhen, or perhaps, as I say, here and now. The inward-looking final remark: what about 'lesser fleas', are the quarks the end or do they not have their own sub-structures? Look into a bacterium and you find it is made of molecules; look into a molecule and you find it is made of atoms; look into an atom and you find it is made of electrons and the nucleus; look into a nucleus and you find it is made of neutrons and protons; look into a neutron or proton and you find it is made of quarks; look into a quark and . . .? I am sure that I need not tell you that speculation is rife. The only argument against such infinite regression is that of Lucretius:

> (in that case), however endlessly infinite the Universe may be, yet the smallest things will equally consist of an infinite number of parts. Since true reason cries out against this and denies that the mind can believe it, you must needs give in and admit that there are least parts which themselves are partless.

Perhaps we should remind Lucretius that: '. . . no fact is so simple that it is not harder to believe than to doubt at the first presentation.'

Amabel Williams-Ellis

The opening chapter from *How you Began*, 1928

You were once a little tiny speck of jelly. This was long before you became a baby or were born. You had no hands, or feet, or arms, or legs, or head.

You had no mouth, or eyes; you were far smaller than a pin's head, quite soft, and no shape in particular.

You did not like anything, or hate anything, or feel glad or sorry.

Only you always meant to grow. Growing was a thing you somehow had to do.

You know when a person is asleep in bed, if the sheet or blanket gets pulled over their nose so that they can't breathe, they wriggle or push it away without waking up.

Well, that was just the way in which you wanted to grow. You had to grow, just as the person who is asleep has to breathe. Wanting to grow is so strong that a plant will move away a stone that stops it from pushing out of the ground,

But for a long time you, the little speck of jelly, went on growing without waking up, just as a person goes on quietly breathing all night. . . .

Once, long ago, there were no proper animals, or fishes, or insects, in the world, except little spots of jelly far smaller than pins' heads, and not nearly so tidy and round. There were no creatures in the world except this kind, which was just like you when you began.

They were called pro-to-zo-a, and they floated about in the water. It must have been very queer, for these little creatures had no eyes, or arms, or legs, or even flippers or suckers. They could not move about properly or even stay still, but just floated where the winds, or the tides, or the streams, took them. If the water was quite still, they could get along a tiny bit by a sort of rolling and stretching, but if the water moved, it carried thousands of these tiny things along. They could do nothing to stop themselves. They could live in salt or fresh water, and they ate little tiny bits of sea weed or pond weed that got broken off the weeds that were growing, or else tiny green plants that floated about just as they did. But they could not see, and they could not swim, so they had to wait till they happened to float near to a scrap of food. When that happened, the tiny jelly creature opened itself anywhere and swallowed the food.

It had no special mouth and no special stomach, because it was like a bit of plasticine that you can dent in anywhere and wrap round a nut or a small marble. It is rather fun to do this: then you will know how the first creatures in the world ate their meals. But the first jelly creature was ever so much smaller than your bit of plasticine, and later on you will find out why it had to be so small.

But, as you will have guessed, very often the protozoa didn't happen to bump into a bit of food. Sometimes the tide or the stream floated them up on to the land and left them there, high and dry on the beach or the bank. Then there was nothing to float in. So when they had eaten anything that they happened to be touching, they got no more food.

Bits of food were often quite close in the water or on the land. If only the protozoa could have seen or smelt a little, and swum or wriggled a little, they could have got the food. But they couldn't.

So then they died.

Heaps and heaps of protozoa died because they couldn't see, or swim, or wriggle, even an inch to get at a crumb of food. It was great waste of protozoa. This waste went on for thousands and thousands of years and it goes on still, for there are plenty of protozoa still alive.

Now this jelly creature, like everything else that is alive, wanted to live and to grow. It did not want, in the way you might want a book, or want a toy, or want to go out. It wanted, as you want to breathe – must breathe – when you are asleep. The protozoa wanted slowly, and wanted all the time, just as you wanted to grow when you were a little jelly creature, like the first live creature that ever was on the earth.

There was one great difference, though, between you and the first live creatures.

It took thousands and millions of protozoa one after the other, living for millions of years, to grow, ever so gradually, little by little, eyes to see, or noses to smell their food with, and to grow flippers or fins to get to it. Some of them never learnt, and so there are plenty of these tiny jelly creatures still in ponds and in the sea. The bother is that they are so small that they are difficult to see. You would have to use a very strong microscope, and that is difficult till you are grown up.

But there are heaps of them in nearly all shallow water, especially in pond water. They are no colour, and really if you look at a frog's egg, it is very like looking at the first creatures down a microscope, except that the frog's egg is neater and rounder.

But the little bit of jelly that was you, seemed to know in a sort of way what it was doing, and you didn't stay plain jelly for more than about a week.

George Wald

An extract from his article in *Proceedings of the National Academy of Sciences*, 1964

We living things are a late outgrowth of the metabolism of our galaxy. The carbon that enters so importantly into our composition was cooked in the remote past in a dying star. From it at lower temperatures nitrogen and oxygen were formed. These, our indispensable elements, were spewed out into space in the exhalations of red giants and such stellar catastrophes as supernovae, there to be mixed with hydrogen, to form eventually the substance of the sun and planets, and ourselves. The waters of ancient seas set the pattern of ions in our blood. The ancient atmospheres molded our metabolism.

We have been told so often and on such tremendous authority as to seem to put it beyond question, that the essence of things must remain forever hidden from us; that we must stand forever outside nature, like children with their noses pressed against the glass, able to look in, but unable to enter. This concept of our origins encourages another view of the matter. We are not looking into the universe from outside. We are looking at it from inside. Its history is our history; its stuff, our stuff. From that realization we can take some assurance that what we see is real.

Judging from our experience upon this planet, such a history, that begins with elementary particles, leads perhaps inevitably toward a strange and moving end: a creature that knows, a science-making animal, that turns back upon the process that generated him and attempts to understand it. Without his like, the universe could be, but not be known, and that is a poor thing.

Surely this is a great part of our dignity as men, that we can know, and that through us matter can know itself; that beginning with protons and electrons, out of the womb of time and the vastness of space, we can begin to understand; that organized as in us, the hydrogen, the carbon, the nitrogen, the oxygen, those 16 to 21 elements, the water, the sunlight – all, having become us, can begin to understand what they are, and how they came to be.

Sources and Acknowledgements

The editor and publishers would like to thank all those who gave permission to use material reprinted in this book.

P. W. Atkins From *The Creation* by P. W. Atkins; copyright © 1981; reprinted with the permission of W. H. Freeman and Company. **Francis Bacon** From *Novum Organum*, 1620. **Margaret A. Boden** 'In search of unicorns'; this article is reprinted by permission of *The Sciences*, 2 East 63rd Street, New York, NY 10021, and is from the September/October 1983 issue. **William Boyd** *Pathology for the Physician*, 1958 edition, p. 219; reprinted by permission of Lea and Febiger. **Vannevar Bush** From *Science is Not Enough*, by Vannevar Bush, pp. 12 and 191; copyright © 1967 by Vannevar Bush; reprinted by permission of William Morrow and Company, Inc. **Nigel Calder** From *Timescale* by Nigel Calder; copyright © 1983 by Nigel Calder; all rights reserved; reprinted by permission of the author, Chatto and Windus Ltd, and Viking Penguin Inc. **Rachel Carson** From *Silent Spring* by Rachel Carson; copyright © 1963 by Rachel L. Carson; reprinted by permission of the Estate of Rachel Carson, and Houghton Mifflin Company. **Lord Cherwell** From *Chemistry and Industry*, 1954, pp. 940–1; reprinted by permission of the editor of *Chemistry and Industry*. **Winston Churchill** From *Thoughts and Adventures*, 1932, by Winston Churchill; reprinted by permission of the Hamlyn Publishing Group Limited. **Paul Colinvaux** From *Why Big Fierce Animals are Rare: An Ecologist's Viewpoint*; copyright © 1978 by Paul A. Colinvaux; excerpt reprinted by permission of Unwin Hyman Limited, and Princeton University Press. **E. J. H. Corner** From *The Life of Plants*, 1964; reprinted by permission of the author and the University of Chicago Press. **George W. Corner** From *The Hormones in Human Reproduction* by George W. Corner; copyright © 1942, rev. 1947, © renewed 1970 by Princeton University Press; excerpt reprinted with permission of Princeton University Press. **Charles Darwin** First extract is from *Journal of Researches into the Natural History and Geology of the Countries Visited during the Voyage of HMS Beagle round the World*, Ward-Lock Co., 1890; second extract is from his *Autobiography*, 1876. **Erasmus Darwin** Lines from *The Temple of Nature*, 1802, Canto I, lines 295–302; *The Botanic Garden*, 1791, Part I, Canto I, lines 289–66. **Richard Dawkins** Chapter 2, 'The replicators', from *The Selfish Gene* by Richard Dawkins (1976), © Oxford University Press 1976; reprinted by permission of Oxford University Press. **Rene Dubos** From

Mirage of Health: Utopias, Progress, and Biological Change; copyright © 1959 by Rene Dubos, © 1987 by Rutgers, The State University; reprinted by permission of Rutgers University Press. **Freeman Dyson** 'The argument from design' from *Disturbing the Universe*, 1979, by Freeman Dyson; copyright © by Freeman J. Dyson; reprinted by permission of Harper & Row, Publishers, Inc. **Otto R. Frisch** 'Energy from nuclei', from *What Little I Remember*, 1979, Cambridge University Press; reprinted by permission of Cambridge University Press. **Francis Galton** An extract from his notes on journeys in Africa, 1874; quoted in D. W. Forrest, *The Life and Works of a Victorian Genius*, Grafton Books. **William Gilbert** From *De Magnete*, first published in 1600; 1958 Dover edition of 1893 translation by P. Fleury Mottelay; reprinted by permission of Dover Publications, Inc. **June Goodfield** An extract from *An Imagined World*, by June Goodfield; reprinted by permission of A. P. Watt Ltd on behalf of June Goodfield. **J. E. Gordon** An extract from *The New Science of Strong Materials* by J. E. Gordon (Pelican Books, 1968, 1976), copyright © J. E. Gordon, 1968, 1976, reproduced by permission of Penguin Books Ltd; an extract from *Structures* by J. E. Gordon (Pelican Books, 1978), copyright © J. E. Gordon, 1978, reproduced by permission of Penguin Books Ltd. and Plenum Publishing Corp. **David Gould** From *The Black and White Medicine Show*, Hamish Hamilton, 1985; reprinted by permission of Hamish Hamilton Ltd. **Stephen Jay Gould** 'Adam's Navel', from *Natural History*, June 1984; with permission from *Natural History*, vol. 93, no. 6; copyright the American Museum of Natural History, 1984. **Richard L. Gregory** 'Blindness, recovery from', from *The Oxford Companion to the Mind*, copyright © Oxford University Press, 1987; reprinted from *The Oxford Companion to the Mind*, edited by Richard Gregory (1987) by permission of Oxford University Press. **J. B. S. Haldane** Extracts from 'Variation within a species' from *The Causes of Evolution*, 1932, are reprinted with permission from Longman; 'Do Continents Move?', from *Science and Everyday Life*, 1941, by permission of the estate of the author and Chatto & Windus; 'Cancer's a Funny Thing' first appeared in the *New Statesman*, 21 February 1964. **Roy Herbert** 'On first encountering sodium benzoate', 1978, first appeared in *New Scientist*, London, the weekly review of science and technology. **A. H. Hill** From 'The ethical dilemma of science', *Nature*, 1952; reprinted by permission from *Nature*, vol. 170, pp. 388–93, copyright © 1952 Macmillan Journals Ltd. **Fred Hoyle** The extract from *The Frontiers of Astronomy*, 1955, is reprinted by permission of Rubinstein Callingham on behalf of the author. **W. H. Hudson** The extract from 'Hints for adder-seekers', from *The Book of a Naturalist*, 1919, is quoted from E. S. Russell, *The Behaviour of Animals*, 1934, published by Edward Arnold. **Nicholas Humphrey** Chapter 2, 'The social function of intellect', from *Consciousness Regained* by Nicholas

Humphrey (1984), © Nicholas Humphrey 1984; reprinted by permission of Oxford University Press. **Julian Huxley** The extract from *New Bottles for Old Wine*, Chatto & Windus, 1959, is reprinted by permission of A. D. Peters & Co. Ltd. **T. H. Huxley** 'On a piece of chalk' is taken from a British Association Lecture, 1868. **Arthur Koestler** 'Cosmic limerick': a personal gift from the author to Dr Bernard Dixon. **Primo Levi** 'Travels with C', from *The Periodic Table*, tr. Raymond Rosenthal, published by Michael Joseph Ltd; reprinted by permission of Michael Joseph Ltd. and Random House Inc. **Alan Lightman** 'Smile' from *Science 85*; the essay also appeared in *A Modern Day Yankee in a Connecticut Court*; reprinted by permission of Alan Lightman and *Science 85*. **Percival Lowell** An extract from the Conclusion to *Mars and its Canals*, published by Macmillan Ltd, 1906. **Gwyn Macfarlane** An extract from *Howard Florey*, Oxford University Press, 1979; reprinted by permission of Watson, Little Limited. **James Clerk Maxwell** Two extracts from his *Scientific Papers*, Cambridge University Press, 1890/1. **Peter Medawar** 'The future of Man', BBC Reith Lecture 1959, published by Methuen & Co., reprinted by permission of Associated Book Publishers (UK) Ltd; 'Is the scientific paper a fraud?', from a BBC radio broadcast talk published by the BBC in 1964, is reprinted by kind permission of Lady Medawar; 'On "The effecting of all things possible"', from *Pluto's Republic* by Peter Medawar (1982), © Peter Medawar 1982, is reprinted by permission of Oxford University Press. **Michael Newman** 'Cloned poem', published in *The Sciences*, 1982, is reprinted by permission of the author. **Isaac Newton** An extract from a letter to Hawes, 1694, from volume III of the *Correspondence*; reprinted by permission of the Royal Society. **Norman Nicholson** An extract from 'Windscale', from *A Local Habitation*, Faber & Faber, 1972; reprinted by permission of David Higham Associates Limited. **Michael Polanyi** A short extract from 'personal knowledge', from *Journeys in Belief*, Routledge & Kegan Paul, 1958; reprinted by permission of Associated Book Publishers (UK) Ltd, and J. C. Polanyi. **John Polkinghorne** 'Perplexities', from *The Quantum World*, is reprinted by permission of Longman. **Richard Proctor** A footnote in 'Astrology', from *Myths and Marvels of Astronomy*, Longman, 1903. **Charles E. Raven** An extract from *The Creator Spirit*, originally published by Martin Hopkinson, 1932; thanks are due to The Bodley Head Ltd and Unwin Hyman Ltd for their assistance in trying to discover the present-day copyright status of this work. **Theodor Roszak** An extract from 'Technocracy's children', from *The Making of a Counter-Culture*, copyright © 1968 by Theodore Roszak; reprinted by permission of Faber & Faber Ltd, and Doubleday, a division of Bantam, Doubleday, Dell Publishing Group Inc. **Miriam Rothschild** 'A liberating bolt from the blue', published in *The Scientist*, July 1987; reprinted with minor amendments by permission of the author. **Richard Rowan-Robinson** Three extracts from *Cosmic Landscape*, Oxford University Press, 1979; reprinted by permission of the author. **Oliver Sacks** 'The President's speech', from *The Man who Mistook his Wife for a Hat*, © 1970, 1981, 1983, 1984, 1985 by Oliver Sacks; reprinted by permission of Summit Books, a Division of Simon & Schuster, Inc., and Gerald Duckworth and Company Ltd. **Charles Sherrington** Two passages from *Man on his Nature*, Cambridge University Press, 1940; reprinted by permission of Cambridge University Press. **John Maynard Smith** Chapter 1, 'The definition of life', from *The Problems of Biology* by John Maynard Smith (1986), © John Maynard Smith 1986; reprinted by permission of Oxford University Press. **George Steiner** 'Has truth a future?', Bronowski Memorial Lecture, 1978, copyright © 1978 by George Steiner; reprinted by permission of the author. **Marie Stopes** A short extract from *Contraception: Its Theory, History and Practice. A Manual for the Medical and Legal Professions*, G. P. Putnam's Sons, 1932; reprinted by permission of The Putnam Publishing Group. **A. Strange** Extracts from 'On the necessity for a permanent commission on state scientific questions', from the *RUSI Journal*, 1871, reprinted by permission of the Royal United Services Institute for Defence Studies. **Lewis Thomas** 'Notes of a biology watcher: to err is human', from *The Medusa and the Snail*, copyright © 1976 by Lewis Thomas. Originally published in the *New England Journal of Medicine*. Reprinted by permission of Viking Penguin, a division of Penguin Books USA, Inc.; 'Biomedical science and human health: The long-range prospect', from *Daedalus*, summer 1977, reprinted by permission of *Daedalus*, Journal of the American Academy of Arts and Sciences, *Discoveries and Interpretations*, vol. 106, no. 3, summer 1977, Boston Mass. **D'Arcy Wentworth Thompson** An extract from the opening chapter of *On Growth and Form*, Cambridge University Press, 1942; reprinted by permission of Cambridge University Press. **Meredith Thring** The 'Scientist Oath', 1971, first appeared in *New Scientist*, London, the weekly review of science and technology. **Colin Tudge** 'The great cuisines are waiting', chapter 7 of *The Famine Business*, Faber & Faber, 1977; reprinted by permission of A. D. Peters & Co. Ltd. **Alfred Russel Wallace** 'The limits of natural selection as applied to Man', from *Contributions to the Theory of Natural Selection*, Macmillan, 1875. **H. G. Wells** The opening paragraph of *The War of the Worlds*, Heinemann, 1898, reprinted by permission of A. P. Watt Ltd on behalf of the Literary Executors of the Estate of H. G. Wells; an extract from *Experiment in Autobiography* by H. G. Wells, copyright 1934 by Herbert George (H. G.) Wells, copyright © renewed 1962 by George Philip Wells and Francis Richard Wells, reprinted by permission of Faber & Faber Ltd, and Little, Brown and Company. **Denys Wilkinson** 'The organization of the universe', by Sir Denys Wilkinson, from *The Nature of*

Matter: Wolfson College Lectures 1980, edited by J. H. Mulvey, 1981, © Wolfson College, Oxford, 1981; reprinted by permission of Oxford University Press. **Amabel Williams-Ellis** The opening chapter of *How you Began*, Gerald Howe, 1928; thanks are due to The Bodley Head Ltd for their assistance in trying to discover the present-day copyright status of this work. **George Wald** The extract from his article in *Proceedings of the National Academy of Sciences*, 1964, is reprinted by permission of Rubinstein Callingham on behalf of the author.

The *Preface* is based on **Bernard Dixon's** article 'Writing with elegance and style': this first appeared in *New Scientist*, London, the weekly review of science and technology, 29 August 1985.

Every effort has been made to identify sources and copyright owners, and to obtain permission to reprint copyright material. The publishers would be glad to receive details of any errors in the above list in order that corrections can be incorporated in any future edition of this book.